# BEYOND TIME AND SPACE
## Fermion Strata & Consistency Conditions for Sequences, Cosmos Spaces and Gambols

Sequence Theory Equation
Three Consistency Conditions
Sequence Theory Basis of Gambol Theory
Gambol View of UST Strata
Gambol Theory Model Quantum Field Theory
Quadruplicated, Quadplexed Quantum Theory
Unified Subluminal–Superluminal Fermion Fields

## Stephen Blaha Ph. D.
## Blaha Research

**MMXXV**

*To Margaret*

## Some Other Books by Stephen Blaha

*SuperCivilizations: Civilizations as Superorganisms* (McMann-Fisher Publishing, Auburn, NH, 2010)

*All the Universe! Faster Than Light Tachyon Quark Starships & Particle Accelerators with the LHC as a Prototype Starship Drive Scientific Edition* (Pingree-Hill Publishing, Auburn, NH, 2011).

*Unification of God Theory and Unified SuperStandard Model THIRD EDITION* (Pingree Hill Publishing, Auburn, NH, 2018).

*The Exact QED Calculation of the Fine Structure Constant Implies ALL 4D Universes have the Same Physics/Life Prospects* (Pingree Hill Publishing, Auburn, NH, 2019).

*Passing Through Nature to Eternity  ProtoCosmos, HyperCosmos, Unified SuperStandard Theory* (Pingree Hill Publishing, Auburn, NH, 2022).

*HyperCosmos Fractionation and Fundamental Reference Frame Based Unification: Particle Inner Space Basis of Parton and Dual Resonance Models* (Pingree Hill Publishing, Auburn, NH, 2022).

*The Cosmic Panorama: ProtoCosmos, HyperCosmos,  Unified SuperStandard Theory (UST) Derivation* (Pingree Hill Publishing, Auburn, NH, 2022).

*God and and Cosmos Theory* (Pingree Hill Publishing, Auburn, NH, 2023).

*Newton's Apple is Now The Fermion* (Pingree Hill Publishing, Auburn, NH, 2023).

*Cosmos Theory: The Sub-Particle Gambol Model* (Pingree Hill Publishing, Auburn, NH, 2023).

*Cosmos-Universe-Particle-Gambol Theory* (Pingree Hill Publishing, Auburn, NH, 2024).

*Fractal Cosmos Curve: Tensor-based Cosmos Theory* (Pingree Hill Publishing, Auburn, NH, 2024).

*The Eternal Form of Cosmos Theory Third Edition  (*Pingree Hill Publishing, Auburn, NH, 2024).

*Fundamental Constants of Cosmos Theory and The Standard Model  (*Pingree Hill Publishing, Auburn, NH, 2024).

*Geometric Cosmos Geometric Universe  (*Pingree Hill Publishing, Auburn, NH, 2024).

*Structure and Dynamics of Cosmos Theory and the Unified SuperStandard Theory* (Pingree Hill Publishing, Auburn, NH, 2024).

*Black Holes, White Holes, and Superluminal Starship*  (Pingree Hill Publishing, Auburn, NH, 2025).

*The Sequence Theory Foundation of Cosmos Theory and Gambol Theory* (Pingree Hill Publishing, Auburn, NH, 2025).

Available on Amazon.com, bn.com, Amazon.co.uk and other international web sites as well as at better bookstores.

# CONTENTS

# FIGURES and TABLES

# Introduction

The unraveling of the series of events that characterize the evolution of a Physical system is difficult if the Physical system has no beginning and no end. Sequence Theory and Cosmos Theory originate before time and spatial coordinates. Therefore in describing these theories we can only view them independently of coordinates. This book takes that view and specifies consistency conditions to determine key time-independent events.

The events that we consider are: the origin of the Cosmos spaces and of a Parent universe, the limitation of Cosmos spaces to ten spaces, the generation of a primary Parent universe, the generation of sub-universes, and the confinement of gambols to within particles.

After describing consistency conditions in detail the book proceeds to describe Gambol Theory within the framework of the four strata of fundamental fermions. Sequence Theory leads to a view of quarks being composed of gambols. For example the quarks of the second stratum (the quarks we know) are composed of 95939 quarks (gambols) of the first stratum. The second stratum u, and other, quarks each contain 95939 confined gambols – first stratum quarks – each with a mass that is 95939 times smaller than the known u quark mass, but with the same set of internal symmetry quantum numbers.

The book then defines a model Gambol quantum field theory. The theory is extended to a Unified SuperStandard Theory (UST) in a Quadruplicated, Quadplexed Quantum Field Theory of fermions with Strong and ElectroWeak interactions. The quadruplication reflects four strata of fermions composed of quadplexed fermions.

The model theory is then extended by effectively combining subluminal and superluminal parts of quadplexed quantum fields into a Super-Bradyon-Tachyon formulation that we call the SBT formulation of quantum fields.

All in all there is a new formulation of UST results and the Sequence-Cosmos-Gambol Theories that is complete. If they are true theories, as the author believes, then we can say we have reached a true theory that is a perfect simulation of Reality.

The book generates the Cosmos from a mathematical point, and proceeds to experimental results with the derivation of the values of the coupling constants and fundamental fermion masses of the UST and the Standard Model. A highly readable, up to date, view of Physics.

# 1. The Panorama of the Cosmos

When we examine the course of the development of human intelligence we see a broad global sweep from primitive to speaking – to writing. A new mechanism for human development has appeared with the complex implementation of Simulation through Artificial Intelligence. By Simulation we mean the description of events, or possible sequences of events, through powerful computation techniques or through intellectual exercises – revelries and dreams.

Simulations offer a means of transcending material events and indulging in thought at superficial or deep levels. Their freedom from material restrictions (even if they simulate material happenings) enables the consideration of informative variations on Reality that deepen understanding and promote further inquiry.

This book, and the previous books on Sequence Theory and Cosmos Theory (and also Gambol Theory), is an endeavor to understand Physical Reality.

They develop a view of Reality that provokes thought and an understanding of the intricacies of events in Reality. We view it as a comprehensive, intertwined description relating aspects of Reality in ways that were not hitherto considered in any great detail.

The development of our view of the Cosmos extends to before 6,000 years ago in the creation myths of early civilizations. From mythic beginnings a transition to a materialistic formulation of Reality began with the PreSocratic Philosophers. From studies of the ultimate components of Reality (earth, air, fire and water) they progressed to an atomic theory that remained dormant until the 19$^{th}$ Century when Chemistry researches supported an atomic theory developed by Boltzmann and others. From there a rapid theoretical and experimental development occurred led by Quantum Theory and Relativity. These advances led to Quantum Field Theory and a more fundamental view of matter as composed of quarks and leptons together with a view of forms of radiation such as photons and gauge bosons.

We are now at a further stage based on the author's Unified SuperStandard Theory (UST) and its extension to Cosmos Theory and Gambol Theory, and more recently Sequence Theory. Our views of matter have progressed to a new stage: atoms, composite fermions, quarks, and now gambols within the stratums of fermions. There are four strata of fermions such as quarks. We experience fermions of the second level as the known quarks and leptons. We now view quarks as having constituents, gambols, which are fermions of the first stratum. Gambols have very small masses – about a factor of 95939 smaller than the known quarks. Just as quarks are confined within nucleons such as protons and neutrons, gambols are confined within quarks at accessible energies.

In addition to a new view of matter Sequence Theory and Cosmos Theory give a structure for a set of 10 spaces, and specify the groups and interactions within each universe defined for each space. The result is a complete theory of the Cosmos.

One way of determining the success of a Physical Theory is to consider consequences: Does the description successfully produce predictions of new properties and features in such a sufficient quantity that the probability of its success is sufficiently high? The determination of the probability of success may be quantitative or may be based on qualitative considerations.

In so doing it raises very interesting questions. At a number of points, events are described that seem to beg interpretation as the result of a decision. "Why do events (changes) appear in a seemingly evolutionary way?" and "When does an event take place?" There seems to be two possible answers: by a decision by "somebody" or due to a causative event. The answer goes beyond Physics.

In our work we have selected an answer – a causative event – in each case. (This choice does not preclude a "somebody." The causative events may be viewed as consequent to the actions of a "somebody.")

We have chosen to view major causative events as occasioned by a Consistency Condition in each case. This book examines consistency conditions for a variety of cases where a major event happens. They specify logical causes for events. The events that we consider are: the origin of the Cosmos spaces and of a Parent universe, the limitation of Cosmos spaces to ten, the generation of a primary Parent universe, the generation of sub-universes, and the confinement of gambols to within particles.

## 1.1 A View of Consistency Conditions in Cosmos Theory and Sequence Theory

Consistency conditions enable theories to be developed without a *deus ex machina* – without an entity causing changes and evolution. *Sequence Theory, Cosmos Theory and Gambol Theory describe the features and evolution of the Cosmos and universes as a framework for understanding Physical Reality.*

This view of the most fundamental Physics enables us to avoid asking "Why", "When" or "Who." The consistency condition for the Cosmos spaces and the Parent universe merely exist. There is no origin of Everything in our view. It exists outside of a time and space, and thus we view it as eternal without beginning or end.

A comparison that illustrates our view is the medieval question "How many angels can stand on the head of a pin?" If an answer were found, then the next questions would be "Who could place them there?", "How did they get there?" and "When did this event happen?" These questions and their answers parallel possible similar questions and answers that might be raised about the consistency condition for the Cosmos spaces and Parent universe. *Our view makes these possible questions irrelevant in our Physical considerations.*

## 1.2 A View of Processes in Cosmos Theory and Sequence Theory

When we consider Physical and Chemical processes in the physical world they usually progress through a series of stages that lead to a result.[1] In our consideration of the Cosmos we view it as having no beginning or end in time. We can view the Cosmos as a Reality unencumbered by external change but supporting change within its Parent universe where time is defined. There is an important difference between the eternal Cosmos process (if one may call it a process) and the processes seen in "everyday" life.

This difference is cause for exiting Physical questions of "who", "when" and "how." These questions may be addressed in Philosophic and Religious inquiries.

## 1.3 Creation

*Physical* Creation in the everyday sense of creation does not exist. There is no past for the Cosmos: no structures, no substance and no dynamics. The Cosmos had no beginning and therefore was not created in the familiar dynamically evolving manner of a Physical system. The Physical creation implicit in the Consistency Condition presented here is merely a description of the onset of a Cosmos and Parent universe that makes sense Physically. It provides an understanding of the features embodied in the Cosmos, and Parent universe, that might be viewed as an initial condition while realizing that there is no precursor Physics.

The evolution of the Parent universe, and universes created within it, takes place in Parent universe time. Consequently it is subject to the laws of Physics. How and when sub-universes are created is a subject for study. The author believes that there is a Quantum Theory for universes in interaction that might lead to the generation of universes. Universes may also be created due to vacuum energy dynamics.

*Consistency Conditions offer a mechanism for transitions and evolution within the Cosmos without the use of a Prime Mover or other mechanism. The Physical processes of the Cosmos are described by this Physical Theory, which provides an understanding of Cosmos evolution.*

The book generates the Cosmos from a mathematical point. The theory proceeds to calculate experimental results including the derivation of the values of the coupling constants and fundamental fermion masses of the UST and the Standard Model.

*A true Physical theory is a perfect simulation of Reality.*

---

[1] Some processes occur instantaneously at critical points. But they are the consequences of precursor processes.

# 2. The Form of Sequence Theory

## 2.1 The Timeless Cosmos Structure[2]

The structure of the Cosmos is specified by a set Cosmos spaces. See Figs. 2.1a and 2.1b for the set of spaces (not including HyperCosmos spaces of the Second Kind or Limos spaces that we described in previous books).

Since the structure of the Cosmos spectrum of spaces does not have substance, its set of spaces may be viewed as a set of dimension arrays. The spectrum of spaces is a conceptual definition without a need for the specification of an origin. In that respect it is analogous to Plato's Theory of Ideals.

The set of Cosmos spaces is timeless and unchangeable. It exists prior to any definition of time. It is beyond any definition of time. The definition of a time is a feature of a universe. Cosmos spaces are independent of universes. They are essentially a blueprint of the Cosmos. The origin of Cosmos structure is beyond Physics in the author's view. The form of the structure is sequential – based on powers of 2. We attribute the sequential structure to our new Sequence Theory – presented previously – that develops the form of spaces and universes through a set of four types of sequences.

The universes that implement a space definition are created entities. We have shown in previous books that an initial Parent universe(s) must exist that contains (possibly chains of) sub-universes. The origin of a Parent universe is problematic. We can say, as we have previously, that the Parent universe(s) must satisfy a certain energy-like consistency condition for it to exist. Beyond that, we can only view its origin as beyond our current endeavors. The dynamics that generates/creates sub-universes also remains to be thoroughly investigated.

We view the "creation" of the Parent universe as the beginning of a *time*. (There can be no time (or dynamics) prior to the existence of a space-time. A space-time only exists within a universe. Thus time begins within a parent universe at its creation point.) The times of the sub-universes, including their creation times, are relative to the parent universe time. "No time before time."

## 2.2 Primal Cosmos Universe

We have shown in previous books that sub-universes may be created within a parent universe based on consistency condition or based on a Quantum Field Theory of universes. Sub-universes have substance – energy, mass and interactions. Sub-universes would appear to be created as a point and subsequently expanding (Big Bang) to a possibly growing size.

---

[2] This chapter , excepting eqs. 2.2c and 2.2d, appeared in Blaha (2025b).

## 2.3 The Sequence Theory Equation and Cosmos Theory

The set of spaces of Cosmos Theory has a sequential form based on powers of 2. We see it as one of a set of four types of sequences of our *Sequence Theory*. We base Sequence Theory on a fundamental equation, somewhat like the Schrödinger equation but with a scaling form – independent of dimension.

The Sequence equation base of Cosmos Theory must exist timelessly. It specifies the structure of the Cosmos – its spectrum of spaces and their contents, which are sets of dimensions.

The role of the fundamental sequence equation is as a description of the Cosmos. It is not Reality. It is a mathematical reflection of Reality that identifies the steps connecting the parts of Reality.

We have found what may be the core equation of Cosmos Theory that connects the spectrum of Cosmos Spaces, and the features of fermions and interactions in our universe – the only universe with which we are familiar. The core equation is purposefully simple.

It has a basis in a two dimension space with no time variable. We view this space and a Parent universe as originating in a *mathematical point.*[3] Through a sequence of fractal-like line segment increments it becomes a one dimension line. Then through a further sequence of increments it undergoes fractal "transformation" to a unit square of two dimensions. Under scaling, the unit square becomes an infinite two dimension space. We call the resulting space the Fractal Cosmic Curve space.[4] The Cosmic Curve is similar in construction to the Hilbert fractal curve.[5]

We specify it by a simple Schrödinger-like equation in coordinates x and y with a $\lambda/r^2$ scaling potential. The scaling potential provides the simplest guarantee of an energy spectrum that is a sequence of powers of a constant.[6]

$$[(\partial^2/\partial x^2 + \partial^2/\partial y^2)/m + \lambda/m^2 r^2 + C)\psi = 0 \tag{2.1}$$

where $\lambda$ is a coupling constant and $r = (x^2 + y^2)^{1/2}$. In polar coordinates

$$x = r\cos(\varphi)$$
$$y = r\sin(\varphi)$$

we find

$$(\partial^2/\partial r^2 + 1/r\,\partial/\partial r + \partial^2/\partial\varphi^2 + \lambda/mr^2 + mC)\psi = 0 \tag{2.1a}$$
$$(\partial^2/\partial\varphi^2)\psi = 0$$

and the solution

$$\psi = \psi(r)\Phi(\varphi) \tag{2.1b}$$

where

---

[3] There is no past for the Cosmos: no structures, substance or dynamics.
[4] This is detailed in Blaha (2024c), *Fractal Cosmic Curve: Tensor-Based CosmosTheory.*
[5] See the Appendix for details.
[6] This equation may be viewed within the context of conformal symmetry due to its local scale invariant potential.

$$\Phi(\varphi) = \exp(ik\varphi) \tag{2.1c}$$

The quantum number k must be a whole integer to avoid irregularities in $\psi(r)$. We consider[7] the case of $\lambda = k^2 + n_\varphi^2$. Substituting and separating in eq. 2.1a we find

$$(\partial^2/\partial r^2 + 1/r\ \partial/\partial r + (\lambda - k^2)\ /mr^2 + mC)\ \psi(r) = 0$$
$$(\partial^2/\partial r^2 + 1/r\ \partial/\partial r + \lambda_{n_\varphi} r^2 + mC)\ \psi(r) = 0 \tag{2.1d}$$

where

$$\lambda_{n_\varphi} = n_\varphi^2/m \tag{2.1e}$$

The eigenvalue spectrum due to eq. 2.1d is

$$C_n = \text{const}\ \exp(-n\pi/(\lambda_{n_\varphi} - \tfrac{1}{4})^{\frac{1}{2}}) \tag{2.2}$$

based on Case's paper[8] where n and $n_\varphi$ are independent quantum numbers. The quantization of the $C_n$ eigenvalue spectrum is based on a requirement of orthogonality of solutions (an observation due to Von Neumann).

Thus we initially have a radial energy series in eq. 2.2. We generalize this series to a double series in n and an angular momentum quantum number $n_\varphi$.

$$C_{n_\varphi n} = c_{n_\varphi} \exp(n\pi/(\lambda_{n_\varphi} - \tfrac{1}{4})^{\frac{1}{2}}) \tag{2.2a}$$

where n is the radial quantum number and $n_\varphi$ is the angular momentum quantum number. The growth factor for each power series sequence is

$$g_{n_\varphi} = \exp(\pi/(\lambda_{n_\varphi} - \tfrac{1}{4})^{\frac{1}{2}}) \tag{2.2b}$$

*The radial quantum number n parameterizes each sequence of "energy" values. The angular momentum quantum number $n_\varphi$, which we define in the next section, specifies the sequence and its constant factor $c_{n_\varphi}$. The index $n_\varphi$ determines the coupling constant $\lambda = \lambda_{n_\varphi}$ in eq. 2.1, for each specific sequence.*

Each space in the set of 10 HyperCosmos spaces (Fig. 2.1a) defines the size of a dimension array $d_{dr}$ of size:

$$d_{dr} = 2^{r+4} \tag{2.2c}$$

using $g_{n_\varphi} = 2$. Introducing the concept of strata for fermions, where each fermion becomes one of four strata fermions, increases the dimension size to

$$d_{\text{strata dr}} = 2^{r+6} \tag{2.2d}$$

---

[7] Blaha (2025b) considers the Bohr Model, which presents a Physical explanation of the quantization of the coupling constant $\lambda$.

[8] K. M. Case, Phys. Rev. **80**, 797 (1950).

## 2.4 A Look at Sequence Theory

Eq. 2.1 defines the basic equation of Sequence Theory. It applies to all sequences. The sequence relations that follow are:

$$\lambda_{n_\varphi} = n_\varphi^2/m \tag{2.3a}$$

$$g_{n_\varphi} = \exp(\pi/(\lambda_{n_\varphi} - \tfrac{1}{4})^{1/2}) \tag{2.3b}$$

$$\lambda_{n_\varphi} = [(\ln(g_{n_\varphi})/\pi]^{-2} + \tfrac{1}{4} \tag{2.3c}$$

$$C_{n_\varphi n} = c_{n_\varphi} \exp(n\pi/(\lambda_{n_\varphi} - \tfrac{1}{4})^{1/2}) \tag{2.3d}$$

with m = 0.7696. The sequences are characterized by eqs. 2.2, which has the initial values of the sequences: $c_{n_\varphi}$. The constant $c_{n_\varphi}$ is dimensionless for sequences of dimensions. It has the dimension of [mass] for sequences of masses. For Cosmos dimension arrays $c_{n_\varphi}$ is dimensionless and has its value set by the number of creation/annihilation operators for a PseudoQuantum fermion wave function.

*The following section shows that the values of the $\lambda_{n_\varphi}$ for the power series sequences of the Cosmos Theory spaces spectrum, the coupling constants, and the fundamental fermion mass sequences have a simple relationship based on their order: $n_\varphi = 4, 2, 1, \tfrac{1}{2}^{1/2}$ and $\tfrac{1}{2}$.*

## 2.5 The Four Sequences

The application of Sequence Theory shows a clear interrelation among the various sequences of Cosmos Theory. See Fig. 2.2, and Blaha (2024l) and (2025b) for details.

## THE HYPERCOSMOS SPACES SPECTRUM

| Blaha Space Number $N = o_s$ | Cayley-Dickson Number $n$ | Cayley Number $Cn$ $d_c$ | Dimension Array column length $d_{cd}=d_{cr}$ | Dimension Array Size $d_{dN}=d_{dr}$ | Space-time-Dimension $r$ | CASe Group $su(2^{r/2},2^{r/2})$ CASe |
|---|---|---|---|---|---|---|
| 0 | 10 | 1024 | 2048 | $2048^2$ | 18 | su(512,512) |
| 1 | 9 | 512 | 1024 | $1024^2$ | 16 | su(256,256) |
| 2 | 8 | 256 | 512 | $512^2$ | 14 | su(128,128) |
| 3 | 7 | 128 | 256 | $256^2$ | 12 | su(64,64) |
| 4 | 6 | 64 | 128 | $128^2$ | 10 | su(32,32) |
| 5 | 5 | 32 | 64 | $64^2$ | 8 | su(16,16) |
| 6 | 4 | 16 | 32 | $32^2$ | 6 | su(8,8) |
| **7** | **3** | **8** | **16** | $\mathbf{16^2}$ | **4** | **su(4,4)** |
| 8 | 2 | 4 | 8 | $8^2$ | 2 | su(2,2) |
| 9 | 1 | 2 | 4 | $4^2$ | 0 | su(1,1) |

Figure 2.1a. The Cosmos Theory HyperCosmos space spectrum. Note: $n_\varphi = 4$. The parameter n is the Cayley-Dickson number. See Blaha (2022c).

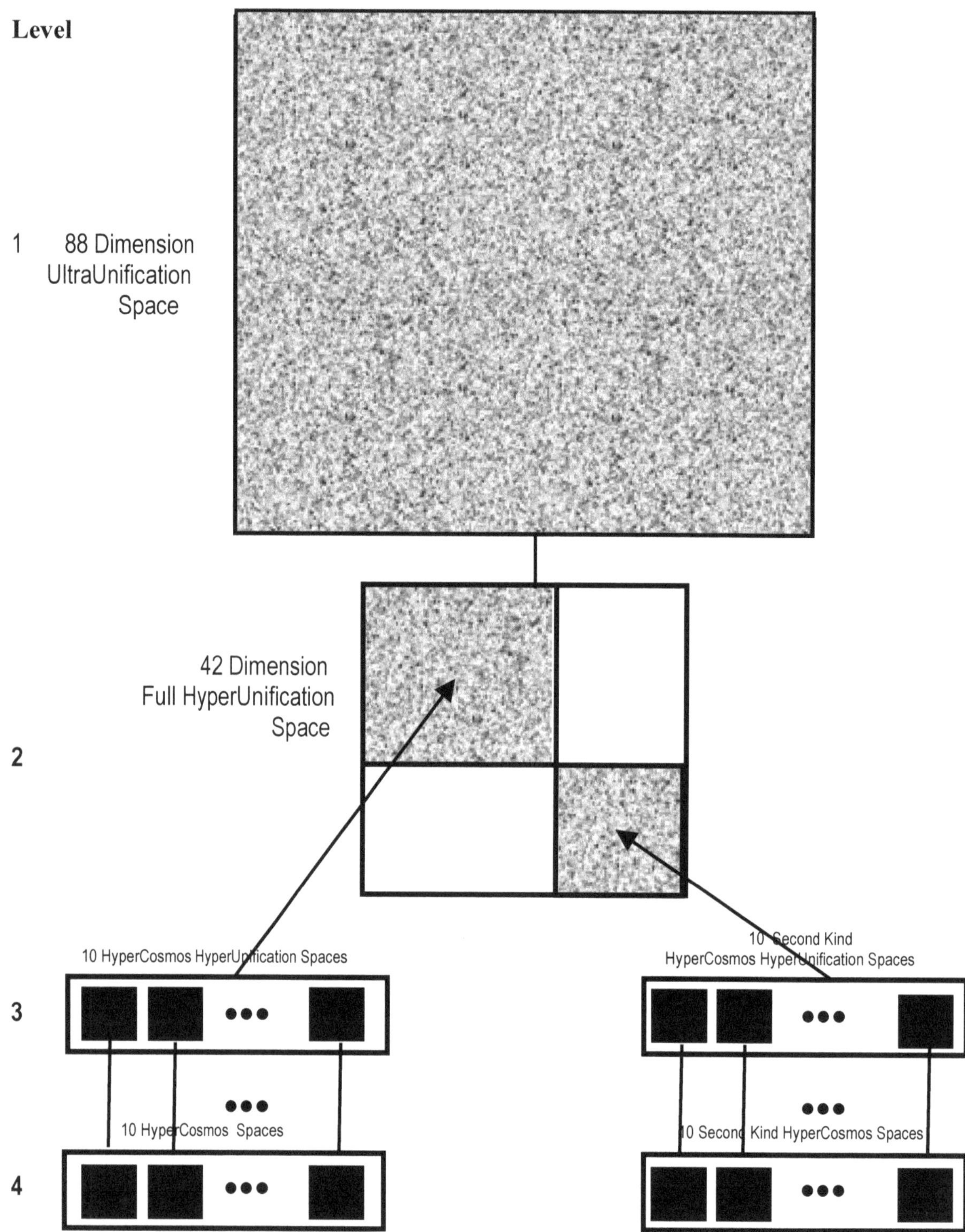


Figure 2.1b. Diagram of the four levels of spaces of Cosmos Theory. It contains 42 spaces. From Blaha (2023a).

| Sequence | $n_\varphi$ | $g_g$<br>multiplier factor | $\lambda_{n\varphi}$ |
|---|---|---|---|
| **A** Cosmos dimension array column lengths $2^{r/2+2}$ | 4 | 2 | 20.79 |
| **B**        Coupling Constants Multiplier | 2 | 4 | 5.39 |
| **C**        Fermion Sequence 2 Masses | 1 | $2/3 \times 2^5$ | 1.3 |
| **D**        Fermion Sequence 1 Masses | $\frac{1}{2}^{\frac{1}{2}}$ | $3/2 \times \pi 2^5$ | 0.65 |
| **E** $n_\varphi$ Values Sequence; A Sequence of Sequences | $\frac{1}{2}$ | 95939 | 0.325 |

Figure 2.2. Interelated Cosmos Theory Sequences. Note: The fermion Mass sequences must apply to mass sequences in other generations and other layers up to overall factors. The coupling constant sequence must apply to the coupling constant sequences in other layers up to overall factors. The values of $n_\varphi$, $g_g$, and $\lambda_{n\varphi}$ should apply for all generations and layers. The sole difference is the values of the starting sequence values $c_{n\varphi n}$. Note $\lambda_{n\varphi} > 0.25$ is required by eq. 2.1 in order to have the bound state energy spectrum. Thus the possible additional line with $n_\varphi = \frac{1}{4}$ is ruled out. Note the multiplier factor increases with $n_\varphi$.

# 3. Consistency Condition for the Number of HyperCosmos Spaces and for a Parent Universe[9]

There is something of a quandary in the definition of a structure for Cosmos spaces and the separate definition/origin of universes. Previously, we have pointed out the importance of a separation of structure (the Cosmos spaces[10] spectrum) from substance (universes).

But the question of their timing (for example, which appeared first?) arises. A space first? Or a time first? Since time is part of a coordinate system, which implicitly or explicitly requires a universe, we found there must be a preexisting universe for time to be defined. However, a universe must be defined based on the definition of a space. Thus we can only conclude the initial *Physical* Cosmos space and universe must be simultaneous in the sense that they cannot be distinguished by a time interval.

Since there is no time before the existence of the universe we cannot specify a creation time. The universe, and Cosmos space, must thus be viewed as without a beginning. Should the universe cease to exist then time ceases as well. Thus we have an eternal initial universe that we call the *Parent*[11] universe. Within it, "sub-universes" may be created according to the time of the Parent universe.

The dimension array of a Cosmos space and its corresponding Parent universe must be simultaneous. The Cosmos space must be one of the set of Cosmos spaces and have a dimension array.

## 3.1 Consistency Condition

The set of Cosmos spaces is initially unlimited. Its dimension r starts at r = 0. It has no apparent upper bound on dimension. We seek a *Physical* basis for an upper bound on dimension to obtain a finite set of Cosmos spaces – a convenient and expedient requirement.

We have suggested a consistency condition for the maximum Cosmos space and Parent universe dimension. The *Physical* principle, which we specified, is to require the Parent universe to not collapse: to be of fixed size or expanding so that it persists eternally. To implement this principle we require a consistency condition that prevents the collapse of the Parent universe. We therefore require the Parent universe be the universe of least dimension (greater than zero) where the virtual thermodynamic outward pressure within the universe is greater than, or equal to, the presumptive

---

[9] See *Fundamental Constants of Cosmos Theory and The Standard Model.* Blaha (2024f). Also Blaha (2022d) and (2023e).

[10] The Cosmos spaces spectrum is defined from the consideration of totally antisymmetric tensors and the creation and annihilation operators of a fermion operator Fourier expansion. It has an equivalent view as a fractal curve that we have constructed and called the Cosmos Fractal Curve.

[11] There could be several "simultaneous" Parent universes.

external Casimir vacuum inward pressure at the time of origin of the Parent universe.[12] See Fig. 3.1. Below this minimum dimension the Parent universe cannot exist since the virtual vacuum pressure would cause it to be contracted into non-existence.

We can summarize the consistency condition as:[13]

*The Parent universe is of the least dimension such that it doesn't contract.*

The dimension r of the Parent universe then fixes a maximum *Physical* Cosmos space dimension. A Parent universe of lesser dimension would collapse. A Parent universe of greater dimension may be stable or collapse. We rule this possibility out to have a fixed minimal dimension Parent universe.

Within the Parent universe there are subuniverses of lower dimension corresponding to lower dimension Cosmos spaces. These universes are not subject to this consistency condition and may expand, be stable or collapse. We discuss their status later.

Recently we suggested a Consistency Condition that specifies the number of HyperCosmos spaces as 10 and also the existence of an 18 dimension Parent universe.

The basis of this analysis was the existence of a Cosmos spaces structure (See Figs. 2.1a and 3.1b.) determined from an analysis of totally asymmetric tensors and the creation/annihilation operators of a fermion field.

We now recapitulate the analysis that leads to the specification of the size of the set of *Physical* Cosmos spaces as numbering 10, and to the dimension of the Parent space to be 18.[14]

## 3.2 Balance Between Expansion and Contraction

When a possible virtual universe is "virtually" created it may expand (Hubble expansion), or it may contract to a point and then cease to exist. The decision is based on the dynamics of the virtual creation. Does the virtual universe possess sufficient mass-energy to generate a pressure for expansion or does the surrounding[15] virtual vacuum pressure cause the universe to contract and be "snuffed out"?

The balance between inward and outward pressure can be represented with

$$H = P - C \tag{3.1}$$

where P is the pressure of the energy-momentum contents of the virtual universe and C is the Casimir force due to the virtual vacuum energy embodied in the surface of the potential universe. There are three possibilities:

---

[12] Much of this chapter is based on Blaha (2022d) and (2023e).

[13] As we see later, the condition has the universe internal pressure rise with dimension. Thus at the equilibrium point the pressures balance. At higher dimensions the universe has higher internal pressure so it would be stable or expand.

[14] We showed previously that it leads to a set of expressions for important constants such as the origin of the power 2 $\cong e\pi/4$; the choice of 10 Cosmos spaces; the choice of a Parent universe of 18 space-time dimensions; and the choice of $c_g = 0.0785 \cong \pi^2/128$, which appears in a universal relation $kT = Mc_g$, as a new universal constant.

[15] In the *surface* of the virtual universe.

$$H < 0 \quad \text{Casimir force would cause contraction to a point} \quad (3.2)$$
$$H = 0 \quad \text{Balance of forces – stable}$$
$$H > 0 \quad \text{Pressure causes (Hubble) expansion of universe}$$

We wish to use H to determine the Cosmos Theory Physical spectrum. The spectrum has a lower Physical limit at r = 0 but no apparent upper limit. We now view H as determining the maximum dimension $r_{Parent}$ of the Physical Cosmos spectrum.

We now proceed to determine H for a universe of dimension r.

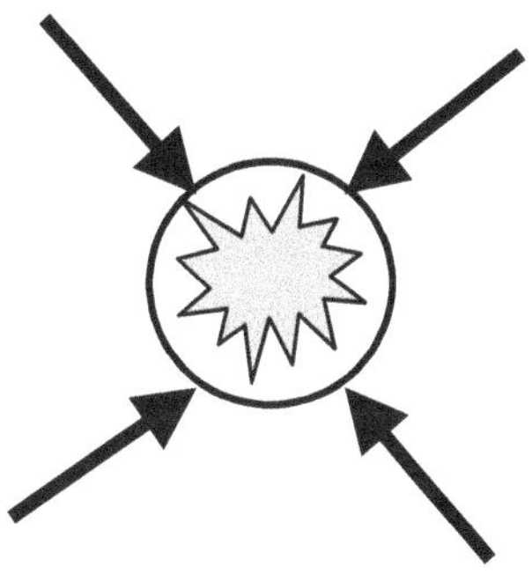

Figure 3.1. Schematic diagram of virtual universe's "vacuum" bubble showing the confining Casimir force of the exterior vacuum. This force is countered by internal fermion gas pressure symbolized by a jagged interior star.

## 3.3 Number of Fundamental Particle Species

Cosmos Theory specifies the number of fundamental fermion species N for a universe in a space of dimension r:

$$N_{Cosmos} = 2^{r+4} \tag{3.3}$$

We assume the number of fundamental fermions is the measure of the number of particles in a universe. The number of fermions can then be used to set P and determine H fixing the minimal universe of dimension r. In Cosmos Theory we have eq. 3.3 specifying the number of fermions for a universe of dimension r. In Sequence Theory we are led to quadruple the number of fermions into strata. See Figs. 3.2 and 3.3. Thus

$$N = S_{equence} = 2^{r+6} \tag{3.4}$$

The argument for quadrupling presented in the previous book for the Unified Super Standard Theory (UST) but applicable to other r universes is:

We define a number $n_s$ labeling a proposed set of replicates of the 256 fermions of the r = 4 Cosmos universe of UST. One replicate of 256 fundamental fermions that includes $v_e$ can be viewed as the $n_s = 1$ set. The next $n_s = 2$ set has the v' neutrino and 255 additional fermions with each of their masses scaled up from their $n_s = 1$ counterpart by a factor of $g_{g^{1/2}}$.

...

The set of strata supports an extension of the fermion Cosmos Theory dimension array. We use them to define a Sequence Theory fermion dimension array of $2^{r+6}$ dimensions due to a factor of $2^2$ from the four stratums. We do not introduce new vector or scalar boson strata for reasons described in chapter 4. This section began the exploration of Sequence Theory.

...

The strata each have a set of fundamental fermions. The fermions of the strata have the same symmetry structure. They differ in mass. The strata masses form sequences based on the multiplier factor $g_{g^{1/2}}$. These masses are modified by quantum field theory effects such as the Higgs Mechanism. Thus their interaction dynamics are affected. Different strata have different dynamics. Yet their group symmetries are the same. *A fermion in one stratum has the same group identity and quantum numbers as the corresponding fermion in a different stratum.*

## 3.4 Conditions at the Consistency Point

We now consider an infinitesimal "box" containing the energy-momentum of a virtual universe. The box contains fermions with the initial features:

1. All fermions are massless.
2. All interactions are zero.
3. All baryon numbers are zero.

## 3.5 Thermodynamic Outward Pressure

We base the thermodynamic pressure on the Ideal Gas Law[16]

$$P = NkT/V \tag{3.5}$$

where P is the pressure, T is the temperature, k is Boltzmann's constant, and N is the number of particles in unit volume V.

Assuming the fermion gas occupies a spherical spatial ball with spatial volume

$$V_{r-1} = \pi^{(r-1)/2} a^{(r-1)} / \Gamma((r-1)/2 + 1) \tag{3.6}$$

with radius a in an r dimension space-time:

$$P = N \, \Gamma((r-1)/2 + 1)kT/( \pi^{(r-1)/2} a^{(r-1)}) \tag{3.7}$$

where $\Gamma(n)$ is the factorial function. The number of fermions N is given by eq. 3.4.

---

[16] One might think of introducing more complex forms of gas laws based on the volume occupied by the gas particles and so on. However our assumption of zero interactions at the origin of universes eliminates the need for such "refinements." One might also thinking of introducing an attractive gravitation term in the consistency condition. Here again the assumption of no interactions at the origin point removes that concern.

| Number of Columns = | 4 NORMAL | 4 | 4 DARK | 4 |
|---|---|---|---|---|
| **Layer 1** 4 generations | e 3 up-quarks | ν 3 down-quarks | e 3 up-quarks | ν 3 down-quarks |
| **Layer 2** 4 generations | e 3 up-quarks | ν 3 down-quarks | e 3 up-quarks | ν 3 down-quarks |
| **Layer 3** 4 generations | e 3 up-quarks | ν 3 down-quarks | e 3 up-quarks | ν 3 down-quarks |
| **Layer 4** 4 generations | e 3 up-quarks | ν 3 down-quarks | e 3 up-quarks | ν 3 down-quarks |

Figure 3.2. The UST 16 $\times$ 16 fermion spectrum of our universe tentatively arranged as SU(4)-plets that correspond directly with SU(4) (or SU(3)$\otimes$U(1)) fermions. There are four layers. Each set of 4 fermions has 4 generations matching the number of rows in each layer. This Periodic Table is broken into Normal and Dark sectors.

**stratum1**

$g_{g^{1/2}} = 95939$

$v_e$   ...

$v_\mu$   ...

$v_\tau$   ...

$v_?$   ...

Plus 3 more layers

**stratum2**

$g_{g^{1/2}} = 95939$

$v_e'$   ... (15 Other fermions)

$v_\mu'$   ... (15 Other fermions)

$v_\tau'$   ... (15 Other fermions)

$v_?'$   ... (15 Other fermions)

Plus 3 more layers

**stratum3**

$g_{g^{1/2}} = 95939$

$v_e''$   ...

$v_\mu''$   ...

$v_\tau''$   ...

$v_?''$   ...

Plus 3 more layers

**stratum4**

$g_{g^{1/2}} = 95939$

$v_e'''$   ...

$v_\mu'''$   ...

$v_\tau'''$   ...

$v_?'''$   ...

Plus 3 more layers

Figure 3.3. A view of the four UST strata of the Sequence Theory fundamental fermions showing mass multipliers. Each stratum has 256 fermions in our universe (in UST theory). Taken together the 4 strata form a $2^{r+6}$ set of fermions in an $r^{th}$ dimension universe. Figure 3.2 shows an SU(4) view of a UST stratum of fermions. Stratum 1 has the known neutrinos. Its charged fermions are the gambols for stratum 2 charged fermions. The stratum 2 charged fermions are the known fundamental fermions with the Dark fermions yet to be found. The 4 strata each contain 4 layers, each of which in turn contains 4 generations of fundamental fermions. The total for the UST is 1024 fermions.

## 3.6 Casimir Effect Confining Force

This section calculates the force on a universe due to the Casimir effect.[17] We assume the universe is in a vacuum energy bubble in an $r - 1$ dimension spatial region of radius a. It generates a confining force on the surface of the universe. We start with the infinite Dirac fermion vacuum energy per unit volume:

$$U_\infty = \int_0^\infty d^{r-1}k \, (k^2 + m^2)^{1/2} \qquad (3.8)$$

where m is the common fermion mass, which we take to be zero by section 3.4.

---

[17] H. G. B. Casimir, Koninkl. Ned. Adak. Wetenschap. Proc. **51**, 793 (1948), S. K. Lamoreaux, Phys. Rev. Lett. **78**, 5 (1997), L. S. Brown and G. J. McClay, Phys. Rev. **184**, 1272 (1969), and references therein.

Due to the vacuum bubble, the vacuum energy changes to

$$U = U_\infty - U_{bubble} \tag{3.9}$$

where the missing bubble vacuum energy per unit volume is

$$U_{bubble} = \int_d^\infty d^{r-1}k \, (k^2 + m^2)^{\frac{1}{2}} \tag{3.10}$$

where d is the momentum space lower limit of the excluded vacuum energy within the bubble. The Casimir deficit is

$$U = \int_0^d d^{r-1}k \, k \tag{3.11}$$

Assuming the momentum d corresponds to the spatial radius, a, of the bubble

$$a = 1/(mc) = 1/d \tag{3.11a}$$

where c is the speed of light we find the energy deficit per unit volume is

$$U \cong \pi^{(r-1)/2} a^{-r} \, (r-1)/[r\Gamma((r-1)/2 + 1)] \tag{3.12}$$

Multiplying by the volume of the bubble we find a total bubble energy:

$$U_{tot} \approx V_{r-1} \, U \tag{3.13}$$

$$\approx \pi^{r-1} a^{-1}[(r-1)/r]/\Gamma((r-1)/2 + 1)^2 \tag{3.14}$$

Then

$$\partial U_{tot}/\partial a \cong -\pi^{r-1} a^{-2} [(r-1)/r]/\Gamma((r-1)/2 + 1)^2 \tag{3.15}$$

A bubble surface of radius a has the area

$$S_{r-2} = 2\pi^{(r-1)/2} \, a^{r-2}/\Gamma((r-1)/2) \tag{3.16}$$

The Casimir force on the surface is therefore

$$\begin{aligned}
C &= (S_{r-2})^{-1} \, \partial U_{tot}/\partial a \\
&= -\tfrac{1}{2} \pi^{(r-1)/2} a^{-r} [(r-1)/r] \, \Gamma((r-1)/2)/\Gamma((r-1)/2 + 1)^2 \\
&= -\pi^{(r-1)/2} a^{-r}/[r\Gamma((r-1)/2 + 1)]
\end{aligned} \tag{3.17}$$

**3.7 H Balance**

We now set the outward pressure of the internal fermion "gas" equal to the inward Casimir vacuum force by eqs. 3.7 and 3.17:

$$H = P - C$$
$$= N\, \Gamma((r-1)/2 + 1)kT/(\pi^{(r-1)/2}a^{(r-1)}) - \pi^{(r-1)/2}\, a^{-r}/[r\Gamma((r-1)/2 + 1)] \qquad (3.18)$$

Eliminating common factors in H we find the condition becomes

$$H' = N\, akT - \pi^{(r-1)}/[r\Gamma((r-1)/2 + 1)^2] \qquad (3.19)$$

At the critical point of eq. 3.2 we see H' = 0 implies

$$N = \pi^{(r-1)}/[arkT\, \Gamma((r-1)/2 + 1)^2] \qquad (3.20)$$

For collapse H' < 0
$$N < \pi^{(r-1)}/[arkT\, \Gamma((r-1)/2 + 1)^2] \qquad (3.21)$$

For expansion H' > 0
$$N > \pi^{(r-1)}/[arkT\, \Gamma((r-1)/2 + 1)^2] \qquad (3.22)$$

**3.8 N and H' as Functions of Dimension r**

We find that the choice of
$$akT = 6.59 \times 10^{-11} \qquad (3.23)$$

and the Cosmos dimension array size choice $N = 2^{r+6}$, give a critical point at dimension r = 18.[18]  Thus at minimum an r = 18 Parent universe can exist.

*For even number dimensions below r = 18 (assuming eq. 3.23) we find H' is negative indicating such a Parent universe could not exist. The Casimir force prevents its existence. At r = 18 the pressure is slightly more (acceptable) then the Casimir force thus allowing an 18 dimension Parent universe within which child universes may be created. The r = 18 virtual universe is the minimal[19] (in r) universe, and thus is the Parent universe dimension.*

On this basis we can choose a Physical set of ten Cosmos spaces. We can create one or more 18 dimension Parent universes. Each Parent universe balances expansion and contraction but favors expansion. A parent universe can contain subuniverses of lesser Cosmos dimension since they have a less strong Casimir force with which to contend.

*If* we wish the Parent universe to be more rapidly *expanding* then we can choose the r = 20 space's universe as the Parent universe. This choice would expand the Physical set of HyperCosmos spaces to eleven.

---

[18] Later in section 3.10 we refine the choice of critical dimension to r = 17.08 based on the critical turning point of the factor $(2e\pi/r)^r$ at r = 17.08 of $2e\pi/r$ from greater than 1 (contraction) to less than 1. A minimum r > 17.08 is required to avoid collapse to a point. We chose the next greater even integer r = 18 as the Parent space dimension. This implies expansion.

[19] We require the dimension to be an even number for reasons stated previously.

## 3.9 Parent Universe Mass-Energy

In earlier work on our Gambol Theory we developed a relationship between internal temperature within a particle and the mass-energy M of gambols within the particle.[20]

$$kT = c_g M$$

where $c_g = 0.0785$, k is Boltzmann's constant and M is the total mass-energy of a particle (the Parent universe in this case).

If we set the radius of the Parent universe to $a = 1/m_0$, then eq. 3.23 gives

$$akT = kT/m_0 = c_g M/m_0$$

implying a very small size $a = 1/m_0$ compared to the mass-energy M of the Parent universe – the Parent universe is very dense:

$$M/m_0 = 8.4 \times 10^{-10}$$

The author's new constant $c_g$ appears in many contexts in our studies in earlier books:

Blaha (2023e) *Cosmos Theory: The Sub-Particle Gambol Model*
Describes its use in deep inelastic lepton and quark resonances, production and interactions, hadron-hadron collisions; neutrino oscillations $e$ - $\mu$, $\mu - \tau$; particle stability pressure vs. Casimir force, and Hubble expansion. See section 9.8 and 9.9. In section 9.9 (eq. 13.22) we find an approximate relation, which we now express as

$$c_g = \pi^2/128 = 0.0771$$

giving a 2% difference from the approximate relation we found $c_g = 0.0785$.

Blaha(2024a) *Cosmos-Universe-Particle-Gambol Theory*
Here we found the following uses of $c_g$:

> The Cosmos Theory Distribution of Universes
> Anti-Planckian Model of Hubble Expansion Parameter
> Our Universe as a Megaverse Particle
> Universe-Megaverse Boundary
> Casimir Force Limit on Hubble Expansion
> $S_8$ Universe-Gambol Clumping Models
> Variation of Universe Cluster Mass-Energy Sizes
> Neutron Star Quark Core Gambol Model
> GVDM Photon – $\rho$ meson Gambol Model
> Higgs and Massive Vector Boson Gambol Models

## 3.10 Support for the Power Series Form of Dimension Arrays

The size of dimension arrays N is a power series in 2 as shown in eq. 3.4. We take eq. 3.20, and use Stirling's approximation for the gamma function; a power series emerges with exponent r for large values of r.

Sterling's approximation is

$$N! = (2\pi n)^{1/2}(n/e)^n \tag{3.24}$$

---

[20] See Gambol Theory in Blaha (2024a).

For large $r \gg 1$, $n \cong r/2$ and

$$\Gamma((r-1)/2 + 1) \rightarrow (\pi r)^{\frac{1}{2}}(r/2e)^{r/2}$$

Substituting in eq. 3.20 gives

$$N = \pi^{(r-1)}/(\text{ark}T) \times 1/(\pi r(r/2e)^r)$$
$$= (2e\pi/r)^r/(\pi \text{ark}T) \tag{3.25}$$
$$= (17.08/r)^r/(\pi \text{ark}T)$$

Or with $N = 2^{r+6}$ we find

$$\text{ak}T = 2^{-r-6}(2e\pi/r)^r/\pi r \tag{3.25a}$$
$$= 2^{-6}e^r\pi^{r-1}/r^{r+1}$$

akT rises with r. For $r = 18$ we have

$$\text{ak}T = 8.47 \times 10^{11}$$

Eq. 3.25 shows the same form of power series with exponent r as N in eq. 3.4. The similarity in form, which is based on the form of volume and surface area in higher dimension spaces, reflects their joint basis in differential forms, n-forms and wedge expressions, as seen before in our earlier books.

*We see a strengthening of the view of dimension array sizes as based on geometry.*

### 3.11 Why Ten Physical HyperCosmos Spaces?

An examination of eq. 3.25 shows that there is a more refined critical point value than $r = 18$, which we set in section 3.8. Due to the factor

$$(2e\pi/r)^r \tag{3.26}$$

The number $2e\pi/r$ changes from greater than one to less than one at $r = 2e\pi = 17.08$. At $r = 18$ the Casimir force has changed from greater than the thermodynamic pressure to less than the thermodynamic pressure OR from contraction to expansion of the Parent universe. Thus $r = 18$ marks the beginning of an acceptable expanding Parent universe. At even integer $r < 18$ universes contract to a point and thus could not generate a Parent universe. The next lowest dimension is even dimension 16 – contraction.

Parent universes can have expanding child universes within them. *Note that the H value of the consistency condition eqs. 3.1 and 3.2 grows from negative to positive as the dimension increases.*

*The key scale for universe, and thus space dimension, is $2e\pi = 8(e\pi/4) \cong 18$. We view this result as a confirmation of the approach in this chapter and the choice of r = 18 as the smallest possible dimension **Physical** Parent Cosmos space.*

### 3.12 Implications of Eq. 3.25

The consistency condition that we have found has a number of important results including the possibility of a new constant of nature $c_g = 0.0785$. This constant appears in many contexts in Elementary Particle theory and Astrophysics.

The implications of the consistency condition are:

1. The $r = 18$ space is the first even dimension integer (most minimal) space whose corresponding universe does not collapse. Thus it is an optimal choice for the Parent space. Within the Parent universe sub-universes may be created that may undergo Hubble expansion. It is possible that there are many such unrelated Parent universes with their own sub-universes.

2. Due to the factor $(2e\pi/r)^r$ the number 17.08 marks a critical dimension for the transition from a contracting universe.

3. The consistency condition embodies a factor $e\pi/4 = 2.13$ where $e = 2.718$ is the natural logarithm base constant. It appears in the factor

$$(2e\pi/r)^r = (8\times(e\pi/4)/r)^r = (8\times(e\pi/4)/r)^r \cong (17.08/r)^r \qquad (3.27)$$

4. Thus $e\pi/4 \cong 2$ is the constant in the dimension array $d_{dr}$ series of powers to within 7% - a good result considering the approximate nature of the consistency condition calculation.

## 3.13 Possible Temperature Dependence on r

If we assume a temperature dependence on the dimension r and set

$$N = 2^{r+6} = (2e\pi/r)^r/(\pi a r k T) \qquad (3.28)$$

we find

$$kT = = 2^{-6}e^r\pi^{r-1}/(ar^{r+1}) \qquad (3.29)$$

## 3.14 A New Level of Physical Reality

The calculation of the Consistency Condition for Cosmos Theory has led to a number of simple expressions based on fundamental mathematical constants that raise the possibility of a new deeper level of Reality at the base of Cosmos Theory and Physics in general:

| Quantity | Expression | Value | Known Value | Deviation |
|---|---|---|---|---|
| $d_{dn}$ Power series Variable | $e\pi/4$ | 2.13 | 2 | 6.5% |
| $c_g = kT/m$ constant | $\pi^2/128$ | 0.0771 | 0.0785 | 1.8% |
| Consistency Condition dimension | $8\ e\pi/4$ | 17.08 | | |

To these values may be added a similar looking expression for $\alpha$:

| | | | | |
|---|---|---|---|---|
| Fine Structure Constant[21] $\alpha$ | $e^2/1024$ | 0.00721438 | 0.0072973525643 | 3.15% |

---

[21] Calculated to known experimental value in the Johnson-Baker-Willey QED by this author. See Blaha (2019f) and references therein. See Blaha (2025b).

where e is the natural logarithm base e = 2.718. The approximation compares extremely well with the experimentally known value of $\alpha$. The Fine structure constant appears as a constant in QED and ElectroWeak theory. These regularities are discussed in Blaha (2025b).

### 3.15. The Chain of Cosmos Development

Cosmos theory as it has evolved is the result of a significant series of stages. In this chapter we list the some of the stages with comments.

## Cosmos Structure[22]                                            Comment

Mathematical Point
Fractal Cosmos Curve
Separation of Curve Line Segments
Map to the Dimensions of Spaces
Separation of Cosmos Two Dimension Grid into Infinitesimal dimension growth
Map of Dimensions to totally Asymmetric Tensors
Map of Total Numbers of these Tensors to Dimension Array Sizes
Spaces – Space-Time and Internal Symmetry dimensions

| | |
|---|---|
| Dimension Structure of spaces based on quadrupling | Assumption based on antisymmetric tensor and n-form structure |
| Choice of certain spaces as structuring universes – 10 spaces | Arbitrary |
| Map of elements of each space to Internal Symmetry group dimensions | Based on Standard Model |
| Choice of some groups to be Connection Groups | To unite fermion sectors |

Map of some elements to space-time dimensions
Unification of Internal Symmetries and Space-time in two composite spaces
Use of a Fundamental Reference Frame to reduce the number of dimensions to one
Result: A comprehensive Structure without Substance

## Cosmos Universes                                               Comment

Substances as Energy-Matter Composites
Universe Structure defined by a space structure
Universes have mass-energy when created
Universes begin as a point particle
Universes begin by expansion (Big Bang)
Universes are particles with a QFT with interactions

| | |
|---|---|
| One universe is the parent of all other universes | May be more than one parent |
| Universes have an internal dynamics based on space symmetries | Space structure determines dynamics |

## Particle Interactions                                          Comment

| | |
|---|---|
| In  our universe, Unified SuperStandard Theory (UST) | consistent with r = 4 space   dimension array |
| Universes of higher dimension spaces – quadrupling versions of UST | assumption |
| Higher dimension particles – quadruplings of UST | assumption |
| Higher dimension life – possible but chemically different | assumption |

---

[22] See the Appendix for details.

# 4. Consistency Condition for a Subuniverse

We considered the Consistency Condition for the maximal appearance of a Cosmos space that would support a Parent universe in chapter 3 and found the r = 18 dimension universe was the lowest dimension universe that would not collapse. Thus the Parent universe dimension is 18 and the set of 10 Cosmos spaces has even-valued dimensions ranging from zero to 18.

We now turn to the question of sub-universes in the Parent universe including sub-universes containing universes.[23] We again consider a Consistency Condition that relates the internal pressure to a vacuum Casimir pressure. A universe may be expanding. But its contents are semi-confined by its Casimir force.

How does it differ from the Consistency Condition of chapter 3? In chapter 3 the internal pressure was assumed to be set by the size of the dimension array (the number of fermions) of the Cosmos space of the universe. The external Casimir force was determined from the "hole" in its *total* vacuum energy (of all possible fermions of all possible universes) due to exclusion from the Parent universe.

In the present case the external Casimir pressure is set by the part of the vacuum energy due to the exclusion of the vacuum energy part of the external fermions (specified by the external universe dimension array) from the "hole" specified by the "child" universe.

This approach is predicated on the assumption that the vacuum of the interior of a universe has a set of fundamental fermions that differs from the set of fundamental fermions of the universe within which it resides. They may, and probably will, have an overlap in their fundamental fermion spectrums. But they are different species due to their residence in different universes.

In Blaha (2023e) we developed a model for gambol confinement within fundamental particles: quarks and leptons. We now apply the same reasoning to develop a model of universes confined within a larger universe. (Our universe is then confined within a Parent universe.) We will show that a Casimir force can confine the constituents, matter and energy, of our universe. The universe also has a statistically based outward force due to the pressure of the mass energy within the universe.

The expansion of our universe may be the result of the combined effect of these forces. This chapter develops the formalism for both forces. Then it considers the point in time when the forces are balanced and the net force for universe expansion is zero. Then the universe may become static or it may proceed to expand further or it may contract. A static universe mirrors an elementary particle reflecting our view that universes are a form of particle.

*We treat our universe as a "bubble" in the Megaverse, our parent universe.*

---

[23] Much of the contents of this chapter appear in the author's book *Cosmos-Universe-Particle-Gambol Theory.*

We assume the universe may be viewed as a set of gambols as we earlier did successfully in a manner similar to the distribution of gambols within a fundamental particle. The universe *gambol temperature* $T_g$ is related to the total mass-energy of the universe as a particle gambol temperature is related to the particle's mass:

$$kT_{gu} = c_g Ms = c_g M/(bt) \tag{4.1}$$

with

$$s = 1/bt \tag{4.1a}$$

where $c_g = 0.0785$, k is Boltzmann's constant, s is the fractionation factor into gambols, $b = 8.67 \times 10^{-19}\ \text{sec}^{-1}$ by earlier gambol studies, and M is the total mass-energy of the universe. We introduced a fractionation factor $s = 1/bt$ that gives the gambol temperature $T_{gu}$ a decreasing value with time t, which we take to be $t = 1/sb$.

At small time t the fractionation is large. At large time t the fractionation declines to a small value.

*We interpret eq. 4.1a as relating sb, treated as an energy-like parameter expression, to a time t. Note sb has the dimension of [inverse time] or equivalently [enrgy]. This relationship is evident from our study of deep inelastic lepton-nucleon scattering in section 4.6 Blaha (2005b) and in references therein.*

## 4.1 Confinement Force on the Child Universe

We assume that gambols may be viewed as "universe-lets" that exist in a universe with its own vacuum. We further assume the parent universe, within which a universe resides, has its own vacuum, which *differs* from the vacuum of the child universe. The universe is treated as a three dimension sphere[24] both in itself and within the parent. The universe and parent universe differ by the number of fundamental fermions: $2^{r+6}$ in the universe and $2^{r'+6}$ in the parent. Based on this difference we find an inward directed (on the universe) Casimir force generated by the *difference between the parent vacuum energy and the child vacuum energy.*

## 4.2 Expansion Pressure of Child Universe Mass-Energy

The contents of the child generate an expansion pressure (force) on the boundary of the child.

$$P = nkT/V \tag{4.2}$$

using the Ideal Gas Law where n is the number[25] of particles in the universe and P is the outward force per unit area. We assume the child universe occupies a spherical volume $V_{universe} = 4/3\ \pi a^3$ with radius a. Thus

---

[24] The time, and other Parent universe, dimensions are not part of the determination of vacuum energies at each instant.

[25] We assume the number of particles is approximated by the number of fundamental fermions specified by the universe's space's dimension array.

$$P = 3nkT/(4\pi a^3) \tag{4.3}$$

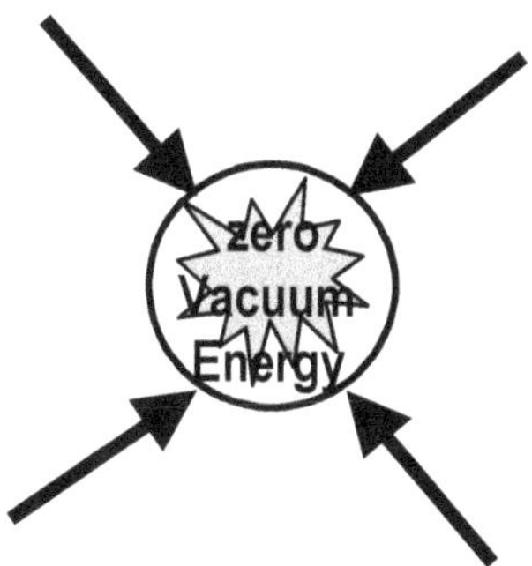


Figure 4.1. Schematic diagram of a universe's vacuum bubble showing the confining Casimir force from the exterior vacuum. This force is countered by internal gambol gas pressure P symbolized by a jagged interior star.

## 4.3 Gambol Model of the Expansion Pressure of Child Universe Mass-Energy

We choose to view the expansion pressure of the child universe in terms of an ideal gas of gambols.[26]

$$P_g = 3nkT_{gu}/(4\pi a^3) \tag{4.4}$$

## 4.4 Energy within a Spherical "Bubble"

The vacuum of the child universe has an energy due to its Dirac sea of fermions. We start with the total infinite Dirac fermion vacuum energy per unit volume summed for the $2^{r+6}$ fundamental fermion species of the child universe.

$$U_\infty = 2^{r+6} \int_0^\infty d^3k \, (k^2 + M^2)^{\frac{1}{2}} \tag{4.5}$$

where M is the average mass.

We now consider a child universe bubble of radius a within the parent, with parent vacuum quantum fluctuations not present, in its Dirac fermion vacuum. The total child vacuum energy per unit volume, using the approximation

$$d \approx 1/a \tag{4.6}$$

is

$$U_{child} = 2^{r+6} \int_d^\infty d^3k \, (k^2 + M^2)^{\frac{1}{2}} = 2^{r+6} \int_{1/a}^\infty d^3k \, (k^2 + M^2)^{\frac{1}{2}} \tag{4.7}$$

or

$$U_{child} = \infty - 2^{r+6}\{4\pi[d(d^2 + M^2)^{3/2}/4 - dM^2(d^2 + M^2)^{\frac{1}{2}}/8 - \text{arsh}(d/(96M^3)]\} \tag{4.8}$$

---

[26] The universe confinement process proceeds analogously for a universe gambol model and a particle model. The difference between the universe and particle models appears when we consider the point in time when the universe pressure equals the Casimir force. Then universe expansion ends and it becomes "stable" like a particle.

per unit volume. The infinite term is not relevant as will be seen when we take the derivative of $U_{child}$. For $d \gg M$ we find (neglecting the infinity)

$$U_{child} \approx -2^{r+6}\pi d^4 \tag{4.9}$$
$$\approx -2^{r+6}\pi a^{-4} \tag{4.10}$$

per unit volume for our universe.

Multiplying by the volume of the bubble we find a total finite vacuum energy for our universe is

$$U_{child} \approx -2^{r+8}\pi^2 a^{-1}/3 = -2^{r+8}\pi^2 d/3 \tag{4.11}$$

**4.5 Vacuum Energy in the Parent Universe**

We find the total bubble energy for the parent is

$$U_{Parent} \approx -4 \times 2^{r'+6}\pi^2 a^{-1}/3 = -2^{r'+8}\pi^2/3a \tag{4.12}$$

*using $2^{r'+6}$ as the number of fundamental fermions in the parent.*

Subtracting the energy of the child universe from the parent energy we find

$$U_{net} = U_{Parent} - U_{child} = -(2^{r'+6} - 2^{r+6})\pi^2/3a \tag{4.13}$$

Eq. 4.13 finds the net energy change due to the difference in energies of the parent and child universes.

**4.6 Casimir Effect Confining Gambol Force**

This section calculates the force on a particle due to the Casimir effect.[27] The inward pressure[28] (Casimir force) on the universe bubble boundary of area, $4\pi a^2$, is

$$P_{Casimir} = -(4\pi a^2)^{-1} \partial U_{net}/\partial a \tag{4.14}$$
$$= (2^{r'+4} - 2^{r+4})\pi/3 \, a^{-4}$$

It is directed inward as shown in Fig. 4.1. It is counterbalanced by gambol thermodynamic pressure. Since the radius a is presumably ultra-small the Casimir pressure will be extremely large. Thus gambols are confined except, perhaps, for "ultimate" energies exceeding the Planck energy.

**4.7 Net Force Gambol Confinement – Universe Pressure**

We now find the difference of the outward pressure of the gambol gas and the inward Casimir vacuum force:

$$F_{net} = = (2^{r'+4} - 2^{r+4})\pi/3 \, a^{-4} - nkT_{gu}/(btV) \tag{4.15}$$

---

[27] H. G. B. Casimir, Koninkl. Ned. Adak. Wetenschap. Proc. **51**, 793 (1948), S. K. Lamoreaux, Phys. Rev. Lett. **78**, 5 (1997), L. S. Brown and G. J. McClay, Phys. Rev. **184**, 1272 (1969), and references therein.
[28] This result is comparable to the often quoted Casimir example of the force between two parallel conducting plates due to the quantum fluctuations of the electromagnetic field between the plates.

where $T_{gu}$ is defined by eqs. 4.1 and 4.4, and where M is the total mass-energy of the universe. Substituting the volume

$$F_{net} = (2^{r'+4} - 2^{r+4})\pi/3 \; a^{-4} - \tfrac{3}{4} \, nc_gM/(\pi bt \, a^3) \qquad (4.16)$$

As in the particle confinement calculation of Blaha (2023e) we set the child particle number with

$$n = 2^{r+6}$$

Consequently

$$\begin{aligned}
F_{net} &= (2^{r'+4} - 2^{r+4})\pi/3 \; a^{-4} - \tfrac{3}{4} \, 2^{r+6}c_gM/(\pi bt \, a^3) \qquad (4.17)\\
&= (2^{r'+4} - 2^{r+4})\pi/3 \; a^{-4} - 2^{r+6}(\tfrac{3}{4} \, c_g/\pi b)M/(ta^3)\\
&= (2^{r'+4} - 2^{r+4})\pi/3 \; a^{-4} - 2.16\times10^{16} \, 2^{r+6}M/(ta^3)
\end{aligned}$$

where $c_g = 0.0785$, k is Boltzmann's constant, $t = 1/sb$ is the time, s is the fractionation into gambols, and $b = 8.67 \times 10^{-19}$ $sec^{-1}$ by earlier gambol studies. M is the total mass-energy of the child universe

## 4.8 When Will the Expansion Force be Zero?

When the time is $t = t_{Zero}$ the Casimir force equals the mass-energy pressure and

$$(2^{r'+4} - 2^{r+4})\pi/3 \; a^{-4} - 2.16\times10^{16} \, 2^{r+6}M/(t_{Zero}a^3) = 0 \qquad (4.18)$$

or

$$\begin{aligned}
t_{Zero} &\approx 2.16\times10^{16} \, 2^{r+6} \, aM \, /((2^{r'+4} - 2^{r+4})\pi/3) \qquad (4.19)\\
&= 2.06\times10^{16} \, 2^{r+6} \, aM \, /(2^{r'+4} - 2^{r+4})\\
&\approx 10^{16} \, 2^{r+7}/(2^{r'+4} - 2^{r+4}) \, aM
\end{aligned}$$

For our 6 dimension parent, where the Megaverse, $r' = 6$ and our 4 dimension universe $r = 4$, we find

$$t_{Zero} \approx 2\times10^{16} \, aM \qquad (4.20)$$

Note: $aM \ll 1$. The value of $t_{Zero}$, the time of creation, is thus within orders of magnitude of the current universe time $t_{NOW} = 4.35 \times 10^{17}$ sec:

$$t_{Zero}/t_{NOW} \approx 0.047 \, aM \qquad (4.21)$$

or $t_{NOW}/t_{Zero} = 21.07/(aM)$.[29] *We thus find the creation time of our universe within the parent Megaverse is much less than the universe's "lifetime." Creation time is measured in terms of our universe's time. The creation time elapsed in the Megaverse as the same amount of time.*

---

[29] Note the interesting relation $\pi e^2 = 23.2 \approx 21.07$ suggesting a deeper connection.

# 5. Consistency Condition for Gambol Confinement in a Particle

In the 1900s major Physicists,[30] such as Abraham, Einstein, Mie, Pauli, Poincare, and Weyl, wrestled with the problem of the interior of fundamental particles – the electron in particular. Their work was based on ElectroDynamics which was the only known force. Other forces such as those of The Standard Model and General Relativity were as yet unknown.

These Physicists found that ElectroDynamics could pretty much account for the internals of electrons. The main sticking point was the need for a confining force for the electron. Such a force was then unknown.

In the absence of a confining force, and with the appearance of Quantum Theory in the nineteen teens to the 1930s followed by the discovery of new particles and new interactions in the 1940s through the 1970s, the program to discover the interiors of fundamental particles was put aside until the present. This author began the search for the interior structure of fundamental particles in Blaha (2022b) and (2022d) using the Limos Sectors mini spaces and the energy structure of the vacuum.

There are several possible "outward pushing" interior forces. One possibility is a Strong Interaction force, which is usually viewed as playing the role of a confining force for quarks (Color confinement) but may play the role of an *interior* "outward pushing" gambol force within particles. Another possible "outward pushing" force is the ElectroWeak force based on SU(2)⊗U(1). A third possibility that we will pursue here is the pressure of gambols bouncing off the boundary of a particle, where the boundary is assumed to be rigid to leading order.   .

A likely possible "inward pushing" confining force is the vacuum – actually the absence of a Dirac vacuum – within particle interiors. We have previously[31] discussed the possibility of vacuum "deserts" within the interior of elementary particles. The "pressure" (actually the Casimir force) of the external vacuum[32] in an elementary particle would cause confinement. If so, then a delicate balance between external vacuum pressure and the internal pressure of quantum fields is required to give structure to elementary particles.

This chapter describes a mechanism for the confinement of gambols to fundamental fermion particles.[33] It describes a gambol confining force due to a Casimir effect that counteracts a pressure due to gambol collisions with the boundary. In previous books we found that a Planckian probability distribution of gambols existed

---

[30] See pp. 184-206 of W.Pauli, *Theory of Relativity*, (Pergamon Press, London, 1967) and references therein.
[31] Blaha (2022d).
[32] The vacuum as an entity was unknown in the early Twentieth Century.
[33] Parts of the chapter appear in Blaha (2022d).

within a fundamental particle. The *gambol temperature* T is related to the mass of the confining particle:[34]

$$kT = c_g m = c_g m \qquad (5.1)$$

where $c_g = 0.0785$, k is Boltzmann's constant, and m is the mass of the particle.

## 5.1 Expansion Pressure of Gambols within a Particle

We assume that gambols may be viewed as small massive, hard particles existing in a non-Dirac vacuumless region devoid of quantum fluctuations, and undergoing motion. Based on this assumption we assume that the gambols generate a pressure for expansion on the "container" of the particle within which they reside. The containment will be assumed to be due to an inward directed Casimir force generated by the external Dirac quantum vacuum. Thus we picture a particle as a region of space without vacuum energy surrounded by a Dirac vacuum containing vacuum energy.

Under these assumptions the pressure generated by the gambols is

$$P = nkT/V \qquad (5.2)$$

using the Ideal Gas Law where n is the number of gambols and P is the outward force per unit area. Assuming the gambol gas occupies a spherical volume $4/3 \, \pi a^3$ with radius a.

$$P = 3nkT/(4\pi a^3) \qquad (5.3)$$

Figure 5.1. Schematic diagram of a particle's empty vacuum bubble showing the confining Casimir force of the exterior vacuum. This force is countered by internal gambol gas pressure P symbolized by a jagged interior star.

## 5.2 Energy within a Spherical "Bubble"

The vacuum of space has an energy due to its Dirac sea of fermions. In previous work[35] we have estimated the energy of the vacuum, and of an "empty" sphere devoid of Dirac sea fermions within the vacuum, to derive the Casimir energy of a particle.

---

[34] Seen in earlier gambol studies.
[35] See chapter 4 of Blaha (2022d).

We start with the infinite Dirac fermion vacuum energy per unit volume for a particle species:

$$U_\infty = \int_0^\infty d^3k\,(k^2 + M^2)^{\frac{1}{2}} \tag{5.3}$$

where M is the particle mass.

We now consider an empty bubble, with quantum fluctuations not present, in the Dirac fermion vacuum. The absence of a quantum vacuum within a fundamental fermion particle bubble has a finite associated energy per unit volume within the bubble.

$$U = U_\infty - U_{\text{bubble}} \tag{5.4}$$

where the total vacuum energy per unit volume is reduced by

$$U_{\text{bubble}} = \int_d^\infty d^3k\,(k^2 + M^2)^{\frac{1}{2}} \tag{5.5}$$

where d is the momentum space lower limit of the excluded vacuum momentum within the particle bubble. We use the approximation

$$d \approx 1/a \tag{5.6}$$

based on the notion that the quantum fluctuation momenta within 1/a are excluded inside the sphere. Thus

$$U = \int_0^d d^3k\,(k^2 + M^2)^{\frac{1}{2}} = \int_0^{1/a} d^3k\,(k^2 + M^2)^{\frac{1}{2}} \tag{5.7}$$

or

$$U = 4\pi[d(d^2 + M^2)^{3/2}/4 - dM^2(d^2 + M^2)^{\frac{1}{2}}/8 - \text{arsh}(d/M)/(96M^2)] \tag{5.8}$$

per unit volume. For $d \gg M$ we find

$$U \approx \pi d^4 \tag{5.9}$$

Assuming the momentum d corresponds to the spatial radius, a, of the bubble $d \cong a^{-1}$ we find

$$U \approx \pi a^{-4} \tag{5.10}$$

per unit volume.

Multiplying by the volume of the bubble we find a total bubble energy;

$$U_{\text{tot}} \approx 4\pi^2 a^{-1}/3 = 4\pi^2 d/3 \tag{5.11}$$

## 5.3 Casimir Effect Confining Gambol *Force*

This section calculates the force on a particle due to the Casimir effect.[36] A fermion particle has no Dirac quantum vacuum within it. Its bubble in Dirac vacuum energy generates a confining force on the particle's gambols. The confinement force must be counterbalanced by the internal pressure of the gambol "gas" to prevent the collapse of the particle to a point.

The inward pressure[37] (force) on the bubble boundary of area, $4\pi a^2$, is

$$P = (4\pi a^2)^{-1}\, \partial U_{tot}/\partial a \tag{5.12}$$
$$= -\pi a^{-4}/3 \tag{5.13}$$

for $a^{-1} \approx d$.

It is directed inward as shown in Fig. 5.1. It is counterbalanced by gambol thermodynamic pressure. Since the radius a is presumably ultra-small the Casimir pressure will be extremely large. Thus gambols are confined except, perhaps, for "ultimate" energies exceeding the Planck energy.

## 5.4 Gambol Confinement Balance

We now set the outward pressure of the gambol gas equal to the inward Casimir vacuum force by combining eq. 5.13 with eq. 5.2:

$$nkT/V = \pi a^{-4}/3 \tag{5.14}$$

$$nkT = 4\pi^2/9a \tag{5.15}$$

Using eq. 5.1 we see

$$a = 4\pi^2/(9nc_g m) \tag{5.16}$$

$$= 55.88 \text{ m}^{-1}/n \tag{5.17}$$

where m is the mass of the particle and $m = nm_g$.

If we choose n = 8 gambols as we have done in previous books then

$$a = 6.98 \text{ m}^{-1} \tag{5.18}$$
$$a = 0.87/m_g \approx 1/m_g \tag{5.19}$$

Using eq. 5.6 we find d equals the gambol mass:

$$d \approx m_g \tag{5.20}$$

If we let $a = 1/m_g$ then, by eq. 5.16, we see

---

[36] H. G. B. Casimir, Koninkl. Ned. Adak. Wetenschap. Proc. **51**, 793 (1948), S. K. Lamoreaux, Phys. Rev. Lett. **78**, 5 (1997), L. S. Brown and G. J. McClay, Phys. Rev. **184**, 1272 (1969), and references therein.
[37] This result is comparable to the often quoted Casimir example of the force between two parallel conducting plates due to the quantum fluctuations of the electromagnetic field between the plates.

$$4\pi^2/(9n^2 c_g) = 1 \qquad\qquad (5.21)$$

implying

$$c_g = 4\pi^2/(9n^2) = 4.39/n^2 = 0.069 \qquad\qquad (5.22)$$

The value of this $c_g$ estimate is 87% of the value determined previously of 0.0785. The approximate calculation here is thus quite accurate.

*The size of a fundamental; fermion is of the order of 1/m. The size of a gambol is of the order of 1/m_g. Thus the gambol confinement model presented here is a physically reasonable approximation.*

# 6. Gambol View of UST Strata

The fundamental fermions of the four strata of the UST are shown in Fig. 3.3. In this chapter we will describe the first (lowest) two strata and relate them to the known fermions and their gambol content. We have tentatively specified a mass multiplier of 95939 between strata. We have defined each stratum as having 256 fermions. The fermions of each stratum have a corresponding fermion in the other strata with the same quantum numbers. They differ only in mass.

## 6.1 Form of Fermion Spectra

Fig. 6.1 shows the first stratum. This stratum has the known neutrinos arranged in 4 generations. The fourth generation is as yet unknown.

**Stratum 1**

|  | Layer 1: | Generation |
|---|---|---|
| $\nu_e$ | e u d s c b t | 1 |
| $\nu_\mu$ | ... | 2 |
| $\nu_\tau$ | ... | 3 |
| $\nu_?$ | ... | 4 |

Plus 3 more layers

Figure 6.1. The first (and lowest mass) UST stratum. It has 4 layers. These fermions will be assumed to be gambols and are *not* the fermions directly observed in nature with the exception of neutrinos. Each layer has 4 generations. Each generation has 8 fermions plus Dark fermions. The first generation of the first layer is depicted in detail. The fermions of the other generations and layers have an analogous form. Only normal gambol fermions are shown above. The set of Dark sector gambol fermions have the same form and are not shown.

We introduce the gambol picture for stratum 1 by identifying the neutrinos as "known" fundamental fermions that do not have gambols within them.[38] The $\nu_?$ neutrino is yet to be found. We identify the other 252 fermions as gambols that are not free but exist only within fermions of the second (and possibly higher) stratum.

Each of the 252 charged stratum 1 fermion gambols are constituents of the corresponding charged fermion of the second stratum.

The second UST stratum is depicted in Fig. 6.2. The neutrinos of the second strata are apparently unconfirmed. We defined the second stratum neutrino $\nu_e'$ as $\nu'$ in our previous books. See Figs. 6.3 and 6.4. The $\nu'/\nu$ ratio that we estimated in previous books was

$$\nu'/\nu = 8.8 \times 10^{-5} \text{ GeV/c}^2/3.878 \times 10^{-11} \text{ GeV/c}^2 \tag{6.1}$$
$$= 2.2691 0^6$$

The ratio is approximately

$$95939 \, \pi e^2 = 2.227 \times 10^6 \tag{6.2}$$

---

[38] This assumption is subject to possible revision.

where $\pi e^2 = 23.21$ and where the factor 95939 is the multiplier $g_{g\frac{1}{2}}$ for the $n_\varphi$ sequence in Blaha (2025b):

$$g_{g\frac{1}{2}} = \exp(\pi/(\lambda_{\text{Bohr } n_\varphi = \frac{1}{2}} - \tfrac{1}{4})^{\frac{1}{2}}) = 95939 \qquad (6.2a)$$

**We now redefine the v' mass, which is in the 2$^{\text{nd}}$ stratum to be**

$$\mathbf{v' = v_e' = 95939\ v = 3.72 \times 10^{-6}\ GeV/c^2}$$

**in order to be consistent with Sequence Theory in Blaha (2025b). Fig. 6.4 shows the revised fermion spectrum.**

We introduce the gambol picture by identifying the strata 1 and 2 neutrinos as fundamental fermions that do not have gambols within them. The $v_?$ neutrino is yet to be found. We identify the other stratum 1 charged fermions as gambols that are not free but exist only within fermions of the second stratum.

Sets of each of the stratum 1 charged fermion gambols are within the *corresponding* charged fermions of the second stratum. Thus the u' quark of stratum 2 contains a set of 95939 u quarks of the first stratum. And so on. Fig. 6.3 symbolizes the map between charged stratum 1 gambol fermions and their containing stratum 2 charged fermions.

**Stratum 2**

| Layer 1: | | Generation |
|---|---|---|
| $v_e'$   e' u' d' s' c' b' t' | | 1 |
| $v_\mu'$       ... | | 2 |
| $v_\tau'$       ... | | 3 |
| $v_?'$       ... | | 4 |

Plus 3 more layers

Figure 6.2. The second (and lowest mass) UST stratum. It also has 4 layers. Each layer has 4 generations. Each generation has 8 fermions plus Dark fermions. The first generation of the first layer is depicted in detail. The fermions of the other generations and layers have an analogous form. Only normal fermions are shown above. The set of Dark sector fermions have the same form and are not shown. The second stratum fermions mirror those of the first stratum with the same quantum numbers but much larger masses due to the multiplier 95939.

## 6.2 The Gambols of Strata

The charged fermions of strata 3 and 4 may contain corresponding charged gambols of strata below their stratum. For example a charged stratum 4 fermion may contain some assemblage of gambols from strata 1, 2, and 3; and a charged stratum 3 fermion may contain some assemblage of gambols from strata 1 and 2. The close correspondence of fermion properties from stratum to stratum makes gamboling possible. The confinement of gambols in fermions in each stratum beyond stratum 1 seems to be required – modulo possible *very* energetic interactions that liberate gambols.

For simplicity, we provisionally assume neutrinos of any stratum are not composed of gambols. Gamboling is restricted to charged fermions.

The UST spectrum that we have created can be directly extended to higher dimension spaces and universes.

The overall form of the strata in the UST displays a remarkable fourfold sequence of symmetry:

$$\begin{array}{ll}\text{4-plets of up-type and down-type fermions} & (6.3)\\ \text{4 generations} \\ \text{4 layers} \\ \text{4 strata}\end{array}$$

*Fourfoldness can be extended to higher dimension spaces and universes.*

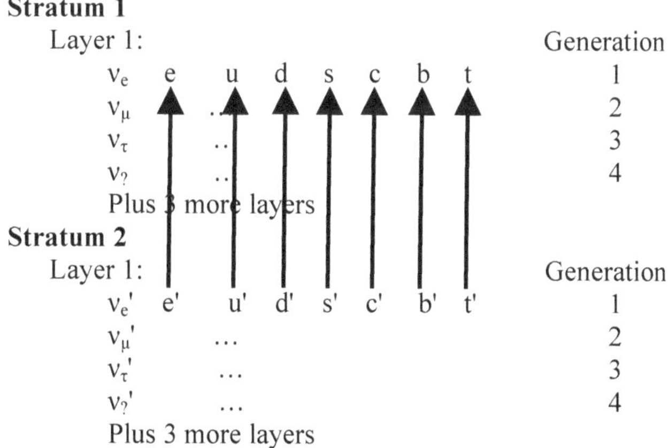


Figure 6.3. Illustration of the map between the first stratum charged gambol fermions and the corresponding second stratum fermions. Only the map between the first generation fermions is displayed. Similar maps are present for the second, third and fourt generations. The Dark sector fermion map is similar but not shown. Only normal fermions are shown above. The set of Dark sector fermions have the same form and are not shown. The second stratum charged fermions mirror those of the first stratum with the same quantum numbers but much larger masses due to the 95939 multiplier. The physical charged fermions are those of stratum 2. They are composed of corresponding charged gambol fermions of stratum 1.

## 6.3 Fermion Quadruplicates

Each fermion of a universe has four quadruplicates. The choice of four quadruplicates of each charged fermion type is based on:[39]

Seeing that there are four sequences above in Fig. 3.4 and that they have a lower limit of the $n_\varphi = \frac{1}{2}$ sequence and an upper limit conceptually[40] of the $n_\varphi = 4$ sequence we suggest that there are 4 counterparts of the 256 fermion spectrum in our universe with corresponding results in other universes of other spaces. We call each counterpart a *stratum* and label them as $n_s = 1, 2, 3, 4$. See Fig. 3.6.

The choice of the quadruplicate number as 4 is additionally based several points of Cosmos spaces' dimension array structure:

---

[39] Relation to sequences taken from Blaha (2025b).
[40] Partitioning dimension arrays further is not significant. Also the $g_g$ is close to 1 making it of less interest.

1. As shown in our earlier work dimension arrays are to symmetry groups and particles as Dirac matrices are to space-times. Both types of arrays are square. Both types of arrays relate directly to dimension through totally antisymmetric tensor properties. Choosing the quadruplication based on 4 we have the augmented dimension array for dimension r of $2^{r+6}$. It is square and is the same as structurally as the dimension array for dimension $r + 2$. As a result we have a consistent formulation. (We note that the augmented r dimension array is only associated with the $2^{r+4}$ set of symmetries. Each stratum's fermions have the same set of symmetries. The $r + 2$ dimension array corresponds to a much larger set of symmetries.)

2. The choice of four replicates for each fermion is analogous to our use of quadplex fermion fields in Blaha (2024$\ell$) to handle ElectroWeak and Strong interactions. Thus there is a consistency in formulation.

The Key Points of quadruplication is:

A. Each set of fermion quadruplicates has identical quantum numbers but differ in mass. When a quadruplicate fermion appears as a gambol within a fermion of a higher stratum its additive quantum numbers are reduced by a factor of 95939. This feature is discussed in detail in chapter 7.

B. The fermions of one stratum are composed of corresponding fermions (gambols) of lower strata (except for the first stratum).

$v_e = 3.878 \times 10^{-11}$ GeV/c$^2$

> Our $v_e$ mass estimate = 0.03878 ev/c$^2$ is consistent with known data estimates.

$v' = \mathbf{95939}\ v = \mathbf{3.72 \times 10^{-6}}$ **GeV/c$^2$**

$e\ = 0.511 \times 10^{-3}$ GeV/c$^2$ = electron mass

$u\ \ 1.76$ mev/c2 = $1.76 \times 10^{-3}$ GeV/c$^2$

> We scale the u mass by a factor of 4/5 – again to put it on a similar footing as the other masses.
> Note: the u mass is 2.2 +0.5 -0.4 $\times 10^{-3}$ GeV/c$^2$ The 1.76 value is consistent with the lower bound 1.8.

$c\ \ 1.27$ GeV/c$^2$

$t\ \ 172.76$ GeV/c$^2$

$d\ \ 3.76 \times 10^{-3}$ GeV/c$^2$

> We scale the d mass by a factor of 4/5 – again to put it on a similar footing as the other masses.
> The 3.76 value is consistent with the tentative lower bound 4.4.

$s\ \ 95 \times 10^{-3}$ GeV/c$^2$

$b\ \ 4.18$ GeV/c$^2$

Figure 6.4. Fundamental first generation current fermion masses. Each fundamental fermion symbol represents its mass. This is a revised Fig. 1.4 in *Quark, Lepton, W and Z Masses of Cosmos Theory and The Standard Model* Blaha (2024g). *Note we are using Sequence Theory to specify the v' mass.*

# 7. Strata, Interactions, and Gambols

This chapter describes features of the Gambol Theory formulation that we have applied previously to particle structure, particle decays, and particle interactions in numerous cases. We begin with a discussion of the gambol construction of a particle. Then we discuss details of gambol interactions. We describe a quadruplicate fermion quantum field theory illustrating its features.

## 7.1 Gambol Construction of a Particle

The manner in which a fermion is composed of gambols is:[41]

1. "A charged fermion of one stratum (except stratum 1) can be treated as composed of a corresponding set of charged fermions (gambols) in the stratum below it. These gambols have identical quantum numbers divided by 95939. Their individual mass is the mass of the fermion divided by the 95939. The largest number of stratum n − 1 gambols in a stratum n fermion  is set by the Sequence Theory multiplier:

$$g_{g\frac{1}{2}} = 95939$$

  The gambols of a fermion are treated as free particles within the fermion.

2. The number of gambols can vary by grouping gambols together in sets. Thus the number of combined gambols could be significantly less than 95939. The fundamental fermions, with which we are familiar, are stratum 2 fermions. The neutrinos  $v_e$, $v_\mu$, and $v_\tau$ are stratum 1 fermions. Stratum 2 fermions can be viewed as an "up to" 95939 gambol assemblage" of stratum 1 gambols. These gambols can be combined into sets down to two gambols for Gambol Theory.

3. For fermions above stratum 2 it is possible to chain together the gamboling: the fermion of a stratum may be treated as an assemblage of next lower stratum fermions (gambols), each of which can be treated as an assemblage of fermions (gambols) of the stratum two stratums below. And so on. As a consequence the number of gambols of lower strata in a charged fermion can be much greater than 95939.

4. As a result a charged fermion of a stratum greater than 1 can consist of a large number of gambols. Each gambol has a fraction of the fermion's mass and additive quantum numbers. It has the spin of the fermion.

5. This procedure may be applied to *hadrons* composed of several quarks by creating a gambol set for each quark and then simply combining the gambols of all quarks within the hadron. See Blaha (2024a) for examples.

---

[41] Adapted from Blaha (2025b).

Example 1:

A fermion of stratum 2 (our known particles) can interact with fermions such as a neutrino of stratum 1 (or of a higher stratum) to generate an interaction like:

stratum 2 fermion $\rightarrow$ other fermions of stratum 1 plus a $v_e$ of stratum 1

Example 2:

A mass M fermion of stratum 2 (our known fermions) can be viewed in Gambol Theory as composed of $g_{g^{1/2}}$ gambols of mass m = M/($g_{g^{1/2}}$). We can then calculate its properties as we did in the Deep Inelastic scattering example considered in section 7.2 below."

The gambol theory applied to Deep-Inelastic Lepton-Nucleon scattering yields a number of key gambol principles that were described in our earlier books and in Blaha (2025b).

## 7.2 Some Gambol Theory Principles

The Gambol Model developed in Blaha (2023e) *Cosmos Theory: The Sub-Particle Gambol Model and* 2023d, *Newton's Apple is Now the Fermion* has numerous implications which we list below. The fractionation process of a particle into sets of gambols is described in Blaha (2023d) and (2023e). Below is a Blaha (2024a) summary with some changes.

1. Fundamental Fermions, Quark and Lepton particles, are composed of probabilistically determined sets of gambols that embody the interior and dynamics of the particle.

2. The probability of each set of gambols within a fermion of mass m is determined by a Planckian distribution:

$$U_{gi}(\varepsilon(s)) = 15\, N(\pi k T_g)^{-4}\, \varepsilon^3/(\exp(\varepsilon/(kT_g)) - 1)$$

where

$$\varepsilon(s) = [(m_g s/m + 1)/(s + 1)]E$$

$$m_g = m/s$$

is the gambol set energy $\varepsilon(s)$ with E the individual gambol dynamic energy, and $m_g$ the gambol mass. Each set of gambols labeled s, initially had s = $2^n$ gambols. We now use s = 95939 or some part of it as the gambol number. Each gambol of a set constitutes a universe of the Limos sector of the Cosmos spaces. $\varepsilon(s)$ interpolates between the fermion mass and the gambol mass. $U_{gi}(\varepsilon(s))$ effectively gives the average number of gambols in a particle for each energy $\varepsilon(s)$. In the case of universes there is a modified $\varepsilon(s)$ designed to have an infinite value at the point of the Big Bang (t = 0).

3. The label s is usually made continuous in calculations.

4. Gambols are confined to the interior of hadrons, fermions and bosons.

5. Gambols are fermions of a lower stratum.

6. The interior of a fermion has a gambol temperature that appears in the fermion's Planckian distribution. The temperature $T_g$ satisfies the law

$$kT_g = 0.0785 \ m$$

where m is the fermion's mass. The constant 0.0785 appears to be related to the confinement of gambols to within fermions.

In the case of universes there is a modified $kT_g$ designed to give $kT_g$ a time dependent value.

7. The total set of gambols in a fermion may be viewed as a gas subject to the laws of Statistical Mechanics and Thermodynamics.

8. Each boson particle acts as an individual gambol.

9. The gambols of a hadron may be viewed as an individually combined composite of the gambols of its individual quarks or as combined into an overall composite for the hadron. The gambol fermions have a fraction of the additive quantum numbers of the enclosing particle. In the simplest case each additive gambol quantum number is reduced by a factor of 95939. The sum of these numbers over all gambols equals the value of the particle quantum number. Similar comments apply to the stratum 2 charged gambols within stratum 3 charged fermions. And similarly for stratum 4 fermions.

For aggregates of gambols the additive quantum numbers of the aggregate gambols add. For a fermion of stratum N whose N – 1 gambols are decomposed into stratum N – 2 gambols the additive numbers aggregate from the N – 2 quantum number values.

10. When a fundamental fermion or hadron interacts it may be viewed as undergoing a gambol interaction with the features:

    a.  The interaction is a sum over sets of gambols probabilistically by individual gambols.

    b.  Each gambol inherits all features of the parent particle except mass, momentum and symmetry quantum numbers which are reduced by 95939 or a factor derived from it due to aggregations or using gambols from different strata.

    c.  The interaction is the lowest order interaction without higher order terms for high energy interactions. We extrapolate this feature to lower energies.

    d.  The S-matrix for the interaction has a Planckian probability factor for each gambol set $U_g(\varepsilon(s))$:

$$S_{fi} = \sum_{\text{Gambols}} \ \prod_{g} U_g(\varepsilon(s)) S_{hadg}$$

where $S_{hadg}$ is the gambol interaction S-matrix factor.

11. Within a fundamental fermion or hadron there is an internal gambol temperature $T_g$ where

$$kT_g = 0.785m$$

where m is the mass of the parent particle and k is Boltzmann's constant. The 0.785 factor is shown to originate in the gambol confinement mechanism described in Blaha (2023e).

12. When a resonance is created or decays the instantaneous gambol temperature of the initiating (or end) particle and the resonance are equal, as are the energies $\varepsilon(s)$ with

$$kT_i = 0.0785\ m_{resonance}$$

13. When hadrons interact, the gambols of each particle interact with those of other particles. *In the interaction region the gambols of all the incoming hadrons are effectively combined in the sense that they are treated as combined for the determination of the structure of the gambol distribution.* The resulting S-matrix is a product of Planckian gambol probabilities and an interaction S-matrix element for interacting gambols of gambol masses and momentums. These terms are summed over the individual gambols.

14. Treating universes as decaying particles of enormous mass-energy, which are composed of gambols, we define a universe gambol temperature with

$$kT_{gu} = 0.0785 m_u/s$$

where s = bt gives a steadily decreasing temperature with time that mirrors the decline of the standard universe temperature to T = 2.72548° K. The constant b = $8.66685 \times 10^{-19}$ sec$^{-1}$.

The above points summarize the gambol models of elementary particles presented in Blaha (2023e). *They help show Sequence Theory is a consistent basis of Gambol Theory.*

## 7.3 Gambol Theory Examples

Numerous examples of gambol applications for deep inelastic lepton-nucleon interactions, particle decay, hadronic decays, and hadronic interactions as well as universe structure may be found in the author's books including

*Cosmos Theory: The Sub-Particle Gambol Model* (Pingree Hill Publishing, Auburn, NH, 2023).

and

*Cosmos-Universe-Particle-Gambol Theory* (Pingree Hill Publishing, Auburn, NH, 2024).

# 8. Gambol Theory Model Quantum Field Theory

We now consider a Quadruplicate fermion Quantum Field Theory model. Corresponding strata fermions all experience the same UST interactions. Corresponding fermions in the strata have the same interactions.[42] Interaction groups are not replicated. A particle of one stratum may interact thereby with particles of other strata.

We will consider the UST set of groups for the various layers and strata. Similar considerations apply to other higher dimension universes of other spaces.

In Fig. 8.1 we see the groups of each stratum are the same for the corresponding layers of each stratum. In layer 1 of each stratum the SU(4) (or SU(3)$\otimes$U(1)) group is the same; the SU(2) $\otimes$U(1) group is the same; and the U(4) groups are the same. *The strata do not have multiples of the symmetry groups and interactions.* We make this choice so that a fermion of one stratum may interact with a fermion of another stratum.

*As a result we may view the fermions of one stratum as composed of the gambols (fermions) of another stratum since their symmetry group features and quantum number structures are the same. This assumption supports the author's Gambol Theory of fundamental fermions. The gambols of one stratum's fermions are the fermions of the stratum above it (excepting stratum one).*

## 8.1 Strata Quantum Field Theory

In this section we will describe a Dirac equation model for the interactions of four fermion strata where each stratum has one fermion.[43] The Lagrangian is

$$\mathcal{L} = \sum_{strata} \overline{\psi}\{i\,\gamma^{\mu}\,\partial/\partial y^{\mu} + \mathcal{S} \sum_{k} g_k A_k{}^{a\mu}(y)\,\gamma_{\mu}T_{ka} + M\}\psi(y) + \text{c.c.} \qquad (8.1)$$

where k labels the interactions. $\psi$ may be viewed as a strata column vector with four components (strata). The k vector boson interaction terms are the same for all four components. We can implement this feature with a Hermitian 4×4 matrix $\mathcal{S}$ where all elements are the integer 1. The mass matrix in this model is a 4×4 matrix for the masses of the four strata. (In this model we treat each stratum as having one fermion. In UST each stratum has 256 fermions.) The free cases of the four fields can each be separately second quantized in the conventional manner with Fourier coefficient operators and conventional anticommutation relations.

The interaction term with the $\mathcal{S}$ matrix may be rewritten as a double sum over strata:

---

<sup></sup>[42] The sources of the fermion masses may be a Higgs Mechanism that differs from stratum to stratum.
[43] The model generalizes directly in the case of our UST universe to 256 fermions per stratum.

$$\sum_{i=1}^{4} \sum_{j=1}^{4} \overline{\psi}_i \sum_k g_k A_k^{a\mu}(y)\, \gamma_\mu T_{ka}\, \psi_j \; + c.c. \tag{8.2}$$

It specifies the interactions are the same for all fermions of all strata. This crucial point of the model is that it shows the interaction term applies equally to all fermions. Thus we can have an interaction with a term

$$\Psi_{n_2} \rightarrow \psi_{n_1} + \gamma \tag{8.3}$$

where $n_1$ and $n_2$ designate strata 1 and 2 respectively, and $\gamma$ represents a vector boson.

The above Lagrangian term make the commonality of the interaction apparent. Each fermion of each stratum experiences the same interaction. We made the interaction explicit above. The interaction could be expressed implicitly as

$$\sum_k g_k A_k^{a\mu}(y)\, \gamma_\mu T_{ka} \tag{8.4}$$

This type of model supports the concept of representing fundamental fermions as assemblages of gambols of a different stratum, as we discussed earlier.

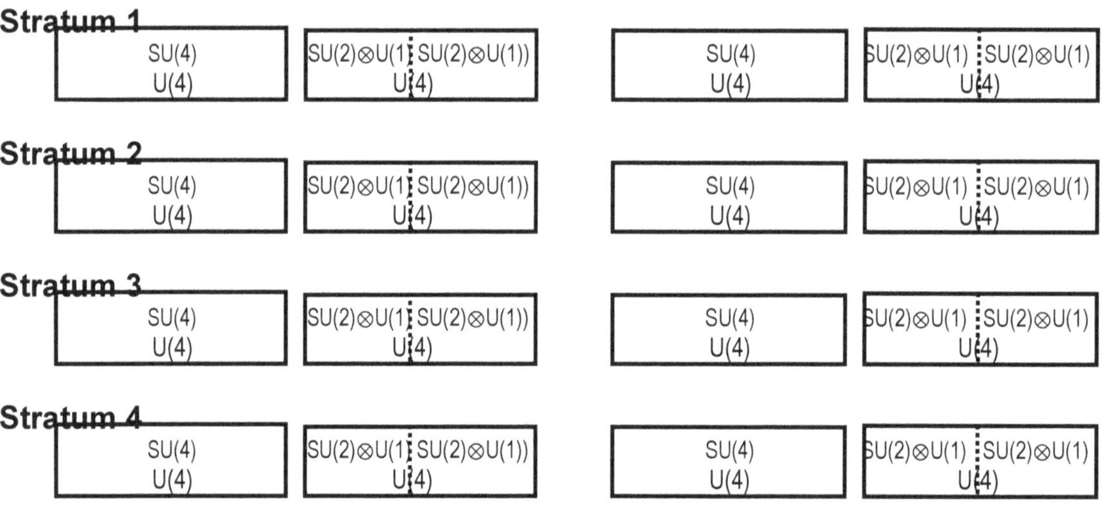


Figure 8.1. **First** layers of the four strata for fundamental fermions. Each stratum has three additional layers with similar relationships between the groups of their layers. Fig. 8.2 below shows the four layers of a single stratum. The group structures are the same for all strata.

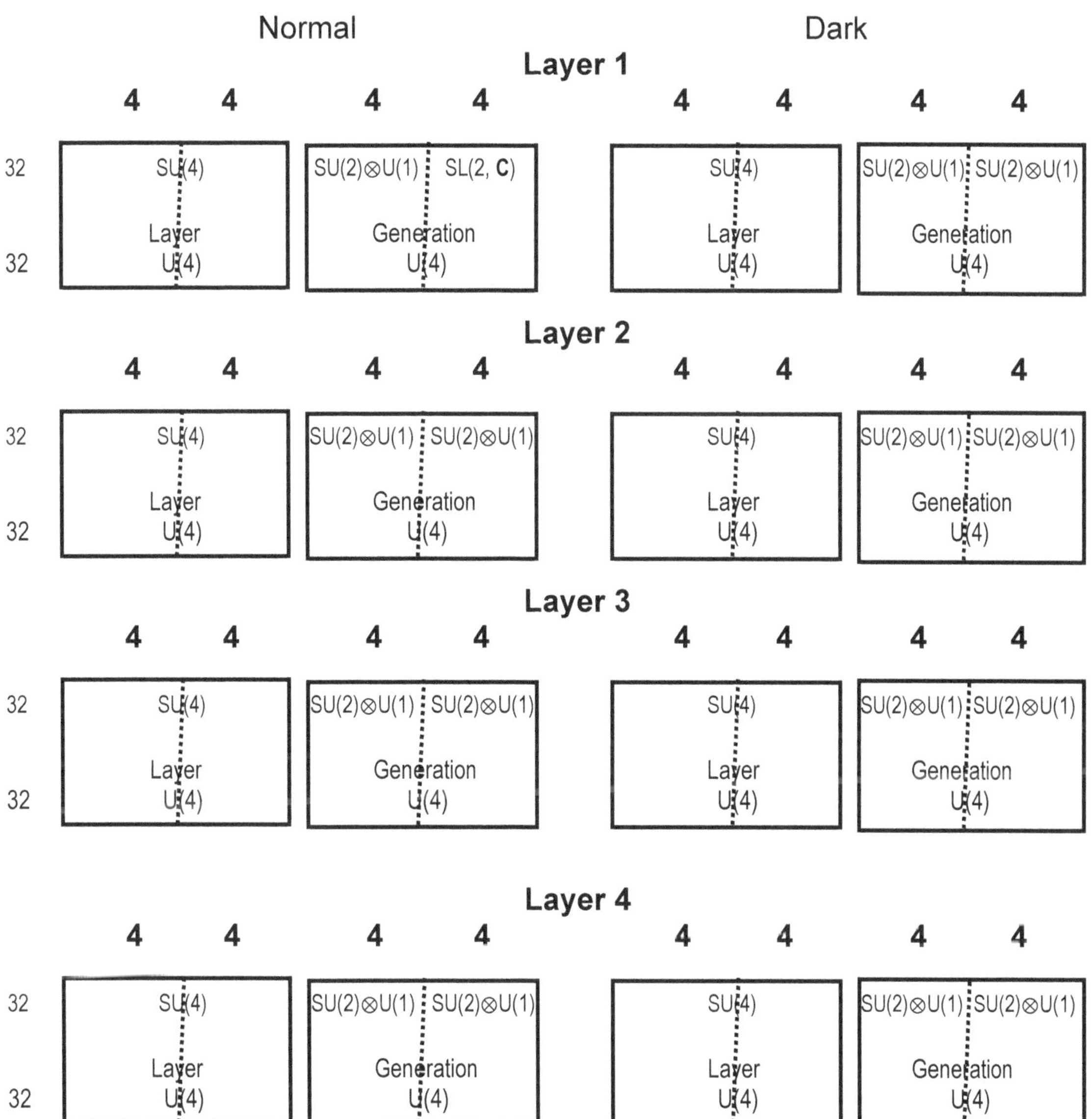


Figure 8.2. (From Blaha (2024i)). Normal and Dark symmetry groups of UST. SL(2, C) represents the Lorentz group SO$^+$(1,3). This diagram appears in Blaha (2024i) and our earlier books such as Blaha (2020d).

# 9. UST Quadruplicated, Quadplexed Quantum Field Theory

A strata's Quadruplicate fermion field formalism can be combined with the Quadplex Fermion fields formalism developed in Blaha (2024j) and (2024ℓ). For each fermion field in the combined formalism there are four replicates – one in each stratum. Each of the four stratum fields of a fermion has four parts, which are bradyon or tachyonic, in the UST. These bradyonic and tachyonic fermion parts are based on the use of four coordinate systems. The fourfold bradyon – tachyon fermion representations are used for implementations of combined Strong SU(4) and ElectroWeak SU(2)⊗U(1) representations. They are described in detail in Blaha (2024j). This chapter extends some of these results.

Sections 9.1 – 9.3 describe the quadplex formulation for the ElectroWeak and Strong SU(4) interactions *for the case of the eight fermions in generation 1 of layer 1 of the UST. These fermions appear in two sequences.* Their interactions are discussed in Blaha (2024j) and (2024ℓ). Section 9.4 describes the formulation when enhanced to include strata of fermions. The quadplex formalism nicely applies to the two fermion sequences using a combination of bradyon and tachyon parts in a composite fermion wave function formulation.

We found that we could develop a representation of ElectroWeak and Strong interactions using four coordinate systems: y and z coordinates for ElectroWeak and u and v coordinates for Strong SU(4).

## 9.1 Quadplex Field Theory

Chapter 6 of Blaha (2024j) developed a Duplex Bradyon-Tachyon (BT) ElectroWeak Theory based on PseudoFermion fields where one (exterior) part corresponded to the LAB[44] with y coordinates and the other (interior) part corresponded to the particle interior with z coordinates. The exterior field was always bradyonic; the interior field was bradyonic for charged leptons and up-type quarks, and tachyonic for neutral leptons and down-type quarks.

A *Quadplex Fermions* formulation can be developed for the Strong SU(4) Interactions starting with an initial SU(4) symmetry group (later broken to SU(3)⊗U(1)) composed of three quarks and a lepton. We again use the distinction between bradyons and tachyons to create a set of four Quadplex fields. Each Quadplex fermion field will consist of four parts where each part is either bradyonic or tachyonic. We associate each quadplex field with a fermion in a fundamental SU(4) representation of four dimensions.

---

[44] LAB coordinates are the ones used directly in experiments. The other coordinates are internal to the dynamics.

We use b to signify a bradyonic field and t to signify a tachyonic field. We will use the up-type and down-type fermion sequences found in our previous books for quadplex fields:

**Up-Type Sequence**

$$e \leftrightarrow \psi_e = \psi_{1\uparrow} \sim b_y b_z t_u t_v \qquad\qquad (9.1)$$
$$u \leftrightarrow \psi_u = \psi_{2\uparrow} \sim b_y t_z t_u b_v$$
$$c \leftrightarrow \psi_c = \psi_{3\uparrow} \sim b_y b_z b_u t_v$$
$$t \leftrightarrow \psi_t = \psi_{4\uparrow} \sim b_y b_z b_u b_v$$

**Down-Type Sequence**

$$\nu_e \leftrightarrow \psi_e = \psi_{1\downarrow} \sim t_y t_z t_u t_v$$
$$\nu' \leftrightarrow \psi_{\nu'} = \psi_{1\downarrow} \sim b_y t_z t_u t_v$$
$$d \leftrightarrow \psi_d = \psi_{2\downarrow} \sim b_y t_z t_u b_v$$
$$s \leftrightarrow \psi_s = \psi_{3\downarrow} \sim b_y t_z b_u t_v$$
$$b \leftrightarrow \psi_b = \psi_{4\downarrow} \sim b_y t_z b_u b_v$$

where we use particle symbols to represent wave functions with coordinates indicated with subscripts. The ElectroWeak part of each fermion uses y and z coordinates, and the SU(4) part uses u and v coordinates (together with y and z parts). All four coordinate systems have four dimensions. The top-type fermion assignment in eq. 9.1 is *generally based on the ordering of quark masses with heavier fermions having more Bradyon parts and lighter fermions having more Tachyon parts.* A similar assignment may be made for down-type fermions. The list is modified to match ElectroWeak pairs with up-type bradyon matched with down-type tachyon except for the e − ν (neutrino) case where the match is e - ν'.

The z, u and v coordinate parts may be bradyonic or tachyonic since these parts are not directly relevant for the Physical LAB frame. The z, u and v coordinates are <u>not</u> "LAB" coordinates. They are internal to the particles.

The y part coordinates are the LAB coordinates. The z, u and v parts are not connected[45] to the LAB coordinates and do not support faster than light motion in the LAB frame.

***The fermion wave functions generally differ. They may have bradyon or tachyon internal parts. As a result the SU(4) generators which are usually represented numerically must be generalized to have bradyon − tachyon projection operator factors as seen in our earlier books. The approach here, based on bradyon-tachyon internal differences, is important for differentiating between fermions. It offers a fermion internal mechanism for understanding SU(4) or SU(3)⊗U(1) interaction experimental data.***

---

[45] Later we will discuss connecting the various parts together with interactions.

## SL(2, C)⊗SU(2)⊗U(1)⊗S(4) Parts

### *9.1.1 SL(2, C)⊗SU(2)⊗U(1)⊗SU(4) Quadplex Parts*

The SL(2, C)⊗SU(2)⊗U(1) part of the UST has 8 real dimensions in the UST dimension array. The SU(4) part of the UST also has 8 real dimensions in the UST dimension array.

The SU(4) structure[46] may be represented by a complex four fermion particle representations according to eq. 9.1. The SU(4) fundamental representation being complex has a part for its four real dimensions (u coordinates) and a part for the four imaginary dimensions (v coordinates). Both parts support bradyon and tachyon sectors. We denote the coordinate system for the real part as u. We denote the coordinate system for the imaginary part as v. The fermion sector combines the SU(4)-specific u and v coordinate systems with the y and z coordinates for the SL(2, C)[47]⊗SU(2) ElectroWeak sector.

Thus the sets of coordinate systems are:[48]

| **SL(2, C)⊗SU(2)** [49] | | **SU(4)** | | (9.2) |
|:---:|:---:|:---:|:---:|:---:|
| y | z | u | v | |
| Lab frame | Imaginary Part | Real Part | Imaginary Part | |
| Bradyon Always | Bradyon or Tachyon | Bradyon or Tachyon | Bradyon or Tachyon | |

### Free PseudoQuantum Lagrangian for All Parts

A free *PseudoQuantum* Lagrangian for the combined symmetries with interaction terms not shown[50] is

$$\mathscr{L} = \overline{\psi}_{\uparrow 2p}{}^{\alpha\beta\rho\sigma}\mathscr{K}_{\uparrow pq}[M_\uparrow{}^{-3}\gamma_{y\alpha\kappa}{}^\mu\partial/\partial y^\mu\ \gamma_{z\beta\lambda}{}^\nu\partial/\partial z^\nu\ \gamma_{u\rho\tau}{}^\mu\partial/\partial u^\mu\ \gamma_{v\sigma\chi}{}^\nu\partial/\partial v^\nu - M_\uparrow]\psi_{\uparrow 1q}{}^{\kappa\lambda\tau\chi} +$$
$$+\ \overline{\psi}_{\uparrow 1p}{}^{\alpha\beta\rho\sigma}\mathscr{K}_{\uparrow pq}[-M_\uparrow{}^{-3}\gamma_{y\alpha\kappa}{}^\mu\partial/\partial y^\mu\ \gamma_{z\beta\lambda}{}^\nu\partial/\partial z^\nu\gamma_{u\rho\tau}{}^\mu\partial/\partial u^\mu\ \gamma_{v\sigma\chi}{}^\nu\partial/\partial v^\nu - M_\uparrow]\psi_{\uparrow 2q}{}^{\kappa\lambda\tau\chi} +$$
$$+\ \overline{\psi}_{\downarrow 2p}{}^{\alpha\beta\rho\sigma}\mathscr{K}_{\downarrow pq}[M_\downarrow{}^{-3}\gamma_{y\alpha\kappa}{}^\mu\partial/\partial y^\mu\ \gamma_{z\beta\lambda}{}^\nu\partial/\partial z^\nu\ \gamma_{u\rho\tau}{}^\mu\partial/\partial u^\mu\ \gamma_{v\sigma\chi}{}^\nu\partial/\partial v^\nu - M_\downarrow]\psi_{\downarrow 1q}{}^{\kappa\lambda\tau\chi} +$$
$$+\ \overline{\psi}_{\downarrow 1p}{}^{\alpha\beta\rho\sigma}\mathscr{K}_{\downarrow pq}[-M_\downarrow{}^{-3}\gamma_{y\alpha\kappa}{}^\mu\partial/\partial y^\mu\ \gamma_{z\beta\lambda}{}^\nu\partial/\partial z^\nu\ \gamma_{u\rho\tau}{}^\mu\partial/\partial u^\mu\ \gamma_{v\sigma\chi}{}^\nu\partial/\partial v^\nu - M_\downarrow]\psi_{\downarrow 2q}{}^{\kappa\lambda\tau\chi}$$

$$(9.3)$$

where $\psi_1$ and $\psi_2$ are PseudoQuantum fields' column vectors containing the various fields in combinations specified in eq. 9.5 below, where p and q designate fermions within sequences according to eq. 9.5 below, where $\gamma_y{}^\mu$, $\gamma_z{}^\mu$, $\gamma_u{}^\mu$ and $\gamma_v{}^\mu$ are Dirac matrices for the y, z, u and v coordinates respectively and $\alpha,\beta,\rho,\sigma,\kappa,\lambda,\tau,\chi$ are spinor indices for quadplex fermions. $M_\uparrow$ and $M_\downarrow$ are the 4×4 mass diagonal matrices of the up-type and down-type fermion sequences respectively.

---

The diagonal matrices $\mathscr{K}_\uparrow^c$ and $\mathscr{K}_\downarrow^c$ handle the numeric factors for the b and t content of the respective individual fermion particles in terms of their bradyon and tachyon content is for generic fermions:

$$\mathscr{K}_\uparrow^c = \begin{bmatrix} (-i)^2 & 0 & 0 & 0 \\ 0 & (-i)^2 & 0 & 0 \\ 0 & 0 & (-i)^3 & 0 \\ 0 & 0 & 0 & (-i)^4 \end{bmatrix} = \begin{bmatrix} -1 & 0 & 0 & 0 \\ 0 & -1 & 0 & 0 \\ 0 & 0 & i & 0 \\ 0 & 0 & 0 & 1 \end{bmatrix} \qquad (9.4)$$

$$\mathscr{K}_\downarrow^c = \begin{bmatrix} 1 & 0 & 0 & 0 \\ 0 & (-i)^2 & 0 & 0 \\ 0 & 0 & (-i)^2 & 0 \\ 0 & 0 & 0 & (-i) \end{bmatrix} = \begin{bmatrix} 1 & 0 & 0 & 0 \\ 0 & -1 & 0 & 0 \\ 0 & 0 & -1 & 0 \\ 0 & 0 & 0 & i \end{bmatrix} \qquad (9.5)$$

### The y and z ElectroWeak Sector Parts

Focusing on the SU(4) related parts, we begin by "factoring" out the y and z dependent parts using the subsidiary equations of motion which have the form:[51]

$$[i\gamma_y{}^\mu \partial/\partial y^\mu - M]\psi = 0 \qquad\qquad (9.6)$$
$$[i\gamma_z{}^\mu \partial/\partial z^\mu - M]\psi = 0 \qquad \text{Bradyonic z field part}$$
$$[\gamma_z{}^\mu \partial/\partial z^\mu - M]\psi = 0 \qquad \text{Tachyonic z field part}$$

where M represents diagonal mass matrices $M_\uparrow$ or $M_\downarrow$ for the up-type and down-type fermion sequences as the case may be here and in the following sections.

### The u and v SU(4) Related Parts for Up-Type Fermions

The u and v parts of the four up-type fermion fields satisfy the subsidiary equations:

$$\begin{aligned} \text{e:} \quad & (T_u T_v)\psi_{T_u T_v} \equiv && [-M^{-1}\gamma_u{}^\mu \partial/\partial u^\mu \, \gamma_v{}^\nu \partial/\partial v^\nu - M]\psi_{T_u T_v} = 0 \\ \text{u:} \quad & (T_u B_v)\psi_{T_u B_v} \equiv && [iM^{-1}\gamma_u{}^\mu \partial/\partial u^\mu \, \gamma_v{}^\nu \partial/\partial v^\nu - M]\psi_{T_u B_v} = 0 \\ \text{c:} \quad & (B_u T_v)\psi_{B_u T_v} \equiv && [iM^{-1}\gamma_u{}^\mu \partial/\partial u^\mu \, \gamma_v{}^\nu \partial/\partial v^\nu - M]\psi_{B_u T_v} = 0 \\ \text{t:} \quad & (B_u B_v)\psi_{B_u B_v} \equiv && [-M^{-1}\gamma_u{}^\mu \partial/\partial u^\mu \, \gamma_v{}^\nu \partial/\partial v^\nu - M]\psi_{B_u B_v} = 0 \end{aligned} \qquad (9.7)$$

interpreting T and B as in equations:

$$B_z\psi = [i\gamma_z{}^\mu \partial/\partial z^\mu - M]\psi = 0 \qquad (4.2c)$$
$$T_z\psi_T = [\gamma_z{}^\mu \partial/\partial z^\mu - M]\psi_T = 0 \qquad (5.2c)$$

---

[51] We use conventional Quantum Field theory from this point in this chapter for convenience. It may be directly generalized to PseudoQuantum Field Theory.

### 9.1.2 Eight Dimension Form of the ElectroWeak and Strong Interactions

The ElectroWeak and the Strong Interactions can be put in an eight dimension matrix form. We define an 8-vector representing the up-type and down-type sequences.

$$\Psi = \begin{bmatrix} e \\ u \\ c \\ t \\ v' \\ d \\ s \\ b \end{bmatrix} \tag{9.8}$$

We define the Strong Interaction SU(4) Lagrangian term:

$$\mathcal{L}_{Strong} = \overline{\Psi} \begin{bmatrix} \begin{array}{c} SU(4) \\ Representation \end{array} & 0 \\ 0 & \begin{array}{c} SU(4) \\ Representation \end{array} \end{bmatrix} \Psi \tag{9.9}$$

We next define the ElectroWeak SU(2)$\otimes$U(1) interaction Lagrangian term:

$$\mathcal{L}_{ElectroWeak} = \overline{\Psi} \begin{bmatrix} \begin{array}{c} SU(2)\otimes U(1) \\ Neutral \\ Vector\ Fields \end{array} & \begin{array}{c} \underline{A^-\ \ 0\ \ 0\ \ 0} \\ SU(2)\otimes U(1) \\ A^+\ Charged \\ Vector\ Fields \end{array} \\ \begin{array}{c} \underline{A^+\ \ 0\ \ 0\ \ 0} \\ SU(2)\otimes U(1) \\ A^-\ Charged \\ Vector\ Fields \end{array} & \begin{array}{c} SU(2)\otimes U(1) \\ Neutral \\ Vector\ Fields \end{array} \end{bmatrix} \Psi \tag{9.10}$$

Note that the electron and neutrino components have charged vector boson factors opposite to the other charged vector boson factors due to their charge (or lack of charge).

We now define the diagonal Lagrangian fermion mass term:

$$\mathscr{L}_{Mass} = \overline{\Psi} \begin{bmatrix} e \\ & u \\ & & c \\ & & & t \\ & & & & v' \\ & & & & & d \\ & & & & & & s \\ & & & & & & & b \end{bmatrix} \Psi \qquad (9.11)$$

## 9.2 Consequences of Subluminal-Superluminal Strong Interaction SU(4)

This section gives a new formulation of The Quadplex Subluminal-Superluminal Standard Model within the framework of UST and the Cosmos Theory. This theory embodies the $SO(1, 3)^{+} \otimes SU(2) \otimes U(1) \otimes S(4)$ part of the first layer of the $r = 4$ dimension array of Cosmos Theory. See Blaha (2024e) for details.

This formulation has several advantages:

1. It explains the apparent faster than light behavior implied by certain fermion spin phenomena in fermion particle models.

2. It opens the possibility of superluminal Physics (and possibly travel) through bradyon-tachyon interactions generating "pure" tachyon (and bradyon) motion. 9.2 UST Quadruplicated, Quadplexed Lagrangian

We now introduce an extension of the quadplex formulation to include four strata of fundamental fermions. A free Quadruplicated, Quadplexed, *PseudoQuantum* Lagrangian for the up-type and down-type fermion sequences for the combined symmetries with interaction terms not shown[52] is the below Lagrangian in eq. 9.12 expressed with a stratum index "a" :

$$\begin{aligned}
\mathscr{L} = \ & \overline{\psi}_{\uparrow 2pa}{}^{\alpha\beta\rho\sigma} \mathscr{K}_{\uparrow pq}[M_{\uparrow a}{}^{-3}\gamma_{y\alpha\kappa}{}^{\mu}\partial/\partial y^{\mu} \, \gamma_{z\beta\lambda}{}^{\nu}\partial/\partial z^{\nu} \, \gamma_{u\rho\tau}{}^{\mu}\partial/\partial u^{\mu} \, \gamma_{v\sigma\chi}{}^{\nu}\partial/\partial v^{\nu} - M_{\uparrow a}]\psi_{\uparrow 1qa}{}^{\kappa\lambda\tau\chi} + \\
+ \ & \overline{\psi}_{\uparrow 1pa}{}^{\alpha\beta\rho\sigma} \mathscr{K}_{\uparrow pq}[-M_{\uparrow a}{}^{-3}\gamma_{y\alpha\kappa}{}^{\mu}\partial/\partial y^{\mu} \, \gamma_{z\beta\lambda}{}^{\nu}\partial/\partial z^{\nu}\gamma_{u\rho\tau}{}^{\mu}\partial/\partial u^{\mu} \, \gamma_{v\sigma\chi}{}^{\nu}\partial/\partial v^{\nu} - M_{\uparrow a}]\psi_{\uparrow 2qa}{}^{\kappa\lambda\tau\chi} + \\
+ \ & \overline{\psi}_{\downarrow 2pa}{}^{\alpha\beta\rho\sigma} \mathscr{K}_{\downarrow pq}[M_{\downarrow a}{}^{-3}\gamma_{y\alpha\kappa}{}^{\mu}\partial/\partial y^{\mu} \, \gamma_{z\beta\lambda}{}^{\nu}\partial/\partial z^{\nu} \, \gamma_{u\rho\tau}{}^{\mu}\partial/\partial u^{\mu} \, \gamma_{v\sigma\chi}{}^{\nu}\partial/\partial v^{\nu} - M_{\downarrow a}]\psi_{\downarrow 1qa}{}^{\kappa\lambda\tau\chi} + \\
+ \ & \overline{\psi}_{\downarrow 1pa}{}^{\alpha\beta\rho\sigma} \mathscr{K}_{\downarrow pq}[-M_{\downarrow a}{}^{-3}\gamma_{y\alpha\kappa}{}^{\mu}\partial/\partial y^{\mu} \, \gamma_{z\beta\lambda}{}^{\nu}\partial/\partial z^{\nu} \, \gamma_{u\rho\tau}{}^{\mu}\partial/\partial u^{\mu} \, \gamma_{v\sigma\chi}{}^{\nu}\partial/\partial v^{\nu} - M_{\downarrow a}]\psi_{\downarrow 2qa}{}^{\kappa\lambda\tau\chi}
\end{aligned}$$
$$(9.12)$$

with quantities defined similarly to those associated with eq. 9.3.

We can express this Lagrangian in a matrix form with 4 component stratum vector fields. Then eq. 9.12 assumes the form of eq. 9.3 with an augmented fermion vector field with 16 components. Eq. 9.8 implies the stratum form:

---

[52] See the reference for the Lagrangian with interaction terms.

$$\Psi = \begin{bmatrix} e \\ u \\ c \\ t \\ v' \\ d \\ s \\ b \end{bmatrix} \qquad (9.13)$$

where each component is a four component vector representing corresponding fermions. Thus "e" designates a *four vector* of fermion fields consisting of the four "e" fermion fields of the "e" strata.

## 9.3 Lagrangian Interaction Terms

The *free* Lagrangian in eq. 9.12 does not have interactions between the fermions within a stratum or from one stratum to another. It is diagonal in fermion quantum fields. Note the mass factors are all completely diagonal matrices in the fermion sequence masses and in the strata. The mass matrices are thus 16×16 diagonal matrices.

The PseudoQuantum SU(4) interaction terms may be obtained by imposing gauge covariance on eq. 9.12. Then one finds the common interaction between the fermions of all strata follow from:

$$\partial/\partial y^\mu \rightarrow \partial/\partial y^\mu - ig\ A^{k\mu}(y)T_{yk}$$
$$\partial/\partial z^\nu \rightarrow \partial/\partial z^\mu - ig\ A^{k\mu}(z)T_{zk} \qquad (9.14)$$
$$\partial/\partial u^\mu \rightarrow \partial/\partial u^\mu - ig\ A^{k\mu}(u)T_{uk}$$
$$\partial/\partial v^\nu \rightarrow \partial/\partial v^\mu - ig\ A^{k\mu}(v)T_{vk}$$

where $T_{yk}$ is the $k^{th}$ SU(4) generator matrix element for the y part of the quadplex fermion field, $A^{k\mu}(y)$ is $k^{th}$ Yang-Mills gauge field for the y coordinates, and similarly for the z, u, and v coordinates. Also g is the coupling constant.

*Note that the interactions are the same for all four strata. These results also apply to the other generations and layers of the UST.*

## 9.4 Expression of fields in a Composite Subluminal-Superluminal Representation

The above use of subluminal and superluminal for the definition of quadplex fermion fields and Lagrangians is somewhat cumbersome. In the next chapter we introduce a new form of quantum field that combines subluminal and superluminal features within a common fermion Fourier expansion. Thus the fields are superficially independent of the distinction between subluminal and superluminal. The differentiation between subluminal and superluminal appears in the fermion states. Fermion states may specify subluminal or superluminal motion. This new approach appears to be more in line with Physical thinking. *The nature of the motion is specified by the states.* The fields are independent of the nature of the motion.

# 10. Unified Subluminal – Superluminal Fermion Quantum Field Theory

In Blaha (2025a) we developed a new solution for White Holes and White Holes of the Second Kind. In addition we examined invariant intervals for subluminal and superluminal motion (for a constant force) and found that the intervals can be specified in the same form with an expression that uses the absolute value of the Lorentz factor

$$|\gamma| = (|1 - v^2/c^2|)^{-\frac{1}{2}} \tag{10.1}$$

The subluminal and superluminal invariant interval expression was found to be:

$$\tau = t/|\gamma| = (mv_c/2F)\,[|\gamma|v/v_c + \mathrm{arsinh}(|\gamma|v/v_c)/|\gamma|] \tag{10.2}$$

where $|\gamma|$ is the absolute value of $\gamma$, F is a constant force, and $v_c$ is the speed of light.

As a result we suggested a covariant superluminal and subluminal force law that extends Newton's Second Law to include superluminal motion:

$$F^\mu = m|\gamma|d(|\gamma|v^\mu)/dt \tag{10.3}$$

In this chapter we apply these considerations to quantum fields. We will define fermion quantum fields that apply equally to subluminal (bradyon) and superluminal (tachyon) motion. The differentiation between these motions appears in the states. Some states are bradyonic. Some states are tachyonic. Some states are a mix of bradyonic and tachyonic. States of quadplex fields are an example of mixed bradyonic/tachyonic parts.

Our purpose is three-fold:

1. To develop a quantum field formalism that unifies subluminal and superluminal motion.

2. To apply this formalism to the duplex and quadplex quantum fields of quadplex ElectroWeak and Strong interaction Theory so as to have a simpler formulation than those presented in chapter 9.

3. To use this new formalism to study transitions between subluminal and superluminal motion at both the quantum and macroscopic classical levels.

We will call the formalism that we develop $\underline{S}$uper-$\underline{B}$radyon-$\underline{T}$achyon (SBT) Quantum Field Theory.

## 10.1 Free SBT Quantum Field Fourier Expansion

In Blaha (2024j) and earlier books we developed bradyon and tachyon quantum field theory. We will now compare bradyon and tachyon quantum field theory Dirac equations and Fourier field expansions with a view to exhibiting their interrelation.[53]

The bradyon Dirac equation is

$$(i\gamma^\mu \partial/\partial x^\mu - m)\psi(x) = 0 \tag{10.4}$$

and the tachyon equation is

$$(\gamma^\mu \partial/\partial x^\mu - m)\psi_T(x) = 0 \tag{10.5}$$

We first note that letting $x^\mu \rightarrow i\, x^\mu$ in eq. 10.4 results in the form of eq. 10.5. If we examine the Fourier expansion of $\psi(x)$

$$\psi(x) = \sum_{\pm s} \int d^3 p N(p)[b(p, s)u(p, s)e^{-ip\cdot x} + d(p, s)v(p, s)e^{+ip\cdot x}] \tag{10.6}$$

and let $x^\mu \rightarrow i\, x^\mu$ we find the exponent $ip\cdot x$ becomes

$$ip\cdot x \rightarrow -p\cdot x = i\,(i\gamma\, p'\cdot x) \tag{10.7}$$

where $p = i\gamma p'$. Noting $\gamma = (1 - v^2/c^2)^{-\frac{1}{2}}$ in eq. 10.6 we see

$$
\begin{aligned}
\mathbf{p} &= m\gamma\mathbf{v} \\
\mathbf{p_T} &= \gamma_T\mathbf{p'} = -i\gamma\mathbf{p'} = -i\mathbf{p} \\
\mathbf{p'} &= m\mathbf{v} \\
p^0 &= \gamma mc^2 = \gamma p^{0\prime} \\
p_T{}^0 &= \gamma_T mc^2 = \gamma_T p^{0\prime} \\
p^\mu &= ip_T{}^\mu
\end{aligned}
\tag{10.8}
$$

where

$$\gamma_T = (v^2/c^2 - 1)^{-\frac{1}{2}} = -i\gamma \tag{10.9}$$

We find

$$ip\cdot x = -p_T\cdot x \rightarrow -i\, p_T\cdot x \tag{10.10}$$

after letting $x^\mu \rightarrow i\, x^\mu$. As a result eq. 10.6 becomes

$$\psi(x) = \sum_{\pm s} \int d^3 p N(p)[b(p, s)u(p, s)e^{-ip_T\cdot x} + d(p, s)v(p, s)e^{+ip_T\cdot x}] \tag{10.11}$$

---

[53] We follow the conventions of Bjorken (1964) and (1965).

achieving the form of the tachyon Fourier expansion. We discuss spinors and creation and annihilation operators later. The tachyon Fourier expansion is

$$\psi_T(x) = \sum_{\pm s} \int d^3p\, N_T(p)[b_T(p, s)u_T(p, s)e^{-ip_T\cdot x} + d_T(p, s)v_T(p, s)e^{+ip_T\cdot x}] \quad (10.12)$$

$\psi_T(x)$ in eq. 10.12 is a tachyon field with tachyon spinors and creation/annihilation operators. (More on these features will be seen below.) We can then write

$$\psi_T(x) = \psi(ix) \quad (10.13)$$

with that understanding. We thus have a map from Dirac bradyon field to a tachyon field.

The change from bradyon (eq. 10.4) to tachyon (eq. 3.5) Dirac equations follows from examining

$$I_\mu = i\, \partial/\partial x^\mu\, e^{-i\gamma p'\cdot x} = \gamma p'_\mu\, e^{-i\gamma p'\cdot x}$$

In the subluminal case it gives

$$I_\mu = p_\mu\, e^{-i\gamma p'\cdot x} = i\, \partial/\partial x^\mu\, e^{-ip\cdot x}$$

corresponding to eq. 10.4. In the superluminal case with $x^\mu \to i\, x^\mu$ we get

$$J_\mu = i\, \partial/\partial x^\mu\, e^{-i\gamma p'\cdot x} \longrightarrow \gamma p'_\mu\, e^{\gamma p'\cdot x} = ip_{T\mu}\, e^{ip_T\cdot x} = \partial/\partial x^\mu\, e^{ip_T\cdot x}$$

by eqs. 10.8 and 10.9 corresponding to the superluminal eq. 10.5. The factor of +i in eq. 10.4 has been replaced by a factor of 1 in eq. 10.5.

The above Fourier expansions use $\int d^3p$. The author developed *light front* coordinate expansions in Blaha (2024j) and earlier books to overcome issues with commutation relations that arise when rectangular coordinates are used.[54] We now turn to an alternate approach.

## 10.2 SBT Quantum Fields

The previous section shows that the difference between bradyon and tachyon quantum fields is due to the difference between the Lorentz Factors for subluminal and superluminal motion. In this respect it parallels the difference for classical motions. We showed subluminal and superluminal motion could be subsumed in the choice of the absolute value of the Lorentz factor (eq. 10.3) in an enhanced Newton's Second Law.

We now implement a similar emendation of fermion quantum fields by using the fermion Dirac equation together with a Fourier expanded field dependent on the absolute value of $\}\gamma|$. The free SBT Dirac equation is[55]

---

[54] The cause of the failure of rectangular coordinate anticommutation relations is the requirement that the magnitude of the 3-momentum is required to be greater than the magnitude of the energy for superluminal motion and thus tachyons.

[55] The SBT equations here can be directly generalized to PseudoQuantum field and to duplex and quadplex fields,

$$(i\gamma^\mu \partial/\partial x^\mu - m)\psi_{SBT}(x) = 0 \qquad (10.14)$$

In the superluminal case it becomes eq. 10.5. The form of the Fourier expansions for both subluminal and superluminal motion is

$$\psi_{SBT}(x) = \sum_{\pm s} \int dV\, N(|\gamma|p')[b(|\gamma|p', s)u(|\gamma|p', s)e^{-i|\gamma|p'\cdot x} + d(|\gamma|p', s)v(|\gamma|p', s)e^{+i|\gamma|p'\cdot x}]$$

$$(10.15)$$

where $p = |\gamma|p'$, is the usual momentum and dV is the integration volume. Eq. 10.15 gives the conventional Fourier expression for subluminal motion. It gives the expansion of eq. 10.12 for superluminal motion. The use of $|\gamma|$ in the Fourier expansion gives eq. 10.14. Effectively, it makes the derivative term appear only negative for superluminal motion in the extended Dirac equation in the superluminal eq. 10.5.

*The use of $|\gamma|$ in eq. 10.15 exactly parallels its use in the author's extension of Newton's Second Law in eq. 10.3.*

Later we see the quadplex extension of eq. 9.12 becomes much simpler using the SBT formalism. For example the $\mathcal{H}_{\uparrow pq}$ factor is eliminated.

### 10.3 Fourier Expansion Integrand $\int dV$

The Fourier expansion of a fermion quantum field such as in eqs. 10.11 and 10.12 leads us to consider a different set of integration parameters. The choice of the momentum **p** creates a difficulty with the equal-time anticommutators, which we described in Blaha (2024j) and earlier books. We now suggest using the magnitude of the velocity v as the alternative to the spatial momentum p.

We now consider the form of Fourier expansions where we transform the $\int dV = \int d^3p$ integration to a $\int dv d\Omega\, J(p,v)$ integration where v is the magnitude of the velocity, and where $J(p,v)$ is its Jacobian. The change to an integral over v enables the integration range to be superficially independent of the nature of the motion whether it is bradyonic or tachyonic.

The conventional subluminal integral in the Fourier expansions uses

$$\int dV = \int d^3p = \int_0^\infty p^2 dp d\Omega \qquad (10.16)$$

where $\int d\Omega$ represents the angular integrations. The integral over p is equivalent to an integral over the magnitude of the velocity v from zero to infinity. It does not include superluminal velocities. By including superluminal velocities we now must use the integral

$$\int dV = \int_0^\infty dv d\Omega\, p(v)^2 J(p,v) \qquad (10.17)$$

The integral over p in eq. 10.16 ranges from m to $\infty$ with a gap from 0 to m. Consequently the equal time anticommutation relations do not give a canonical $\delta^3(r - r')$ result. Using an integral over v eliminates the gap since v ranges from 0 to $\infty$. There is no gap in v. However the choice of v does lead to an algebraic branch point at $v = c$. The point at $v = \infty$ may be taken to be another branch point. Together they define a branch cut from $v = c$ to $v = \infty$.

The Jacobian for the transformation from p to v is

$$J(|p|, v) = dp/dv = m|\gamma| + m(|\gamma|^6 - |\gamma|)^{\frac{1}{2}} \, v/c \qquad (10.18)$$

using

$$v = c(1 - |\gamma|^{-2})^{\frac{1}{2}}$$
$$dv/d|\gamma| = \tfrac{1}{2}\, c(1 - |\gamma|^{-2})^{-\frac{1}{2}}\, (2\,|\gamma|^{-3})$$
$$= c(|\gamma|^6 - |\gamma|)^{-\frac{1}{2}}$$

The Fourier expansion of $\psi(x)$ extended to the velocities from 0 to $\infty$ is:

$$\psi_{SBT}(x) = \sum_s \int_0^\infty dv\, d\Omega\, |p(v)|^2 J(|p|,v) N(|p|)[b(|p|, s)u(|p|, s)e^{-i|p|\cdot x} + d(|p|, s)v(|p|, s)e^{+i|p|\cdot x}]$$
$$(10.19)$$

Note the integral extends from 0 to $\infty$ and thus encompasses both subluminal and superluminal velocities.

## 10.4 SBT Equal Time Anticommutation Relations

The superluminal part of $\psi_{SBT}(x)$ has superluminal spinors and creation/annihilation operators. To have canonical equal time anticommutation relations it is still necessary to use a light front coordinate system. As we said in 2006:

*The only approach that does succeed[56] is to decompose the tachyon field into left-handed and right-handed parts and then second quantize in light-front coordinates.*

The light front representation of $\psi_{SBT}(x)$ quantum behavior is described in detail in chapter 3 of Blaha (2024j) and in (2006) and (2018e). In the light front approach we find: the free, "+" light-front, left-handed tachyon wave field creation/annihilation operators satisfy:

$$\{b_{TL}^{+}(q,s),\, b_{TL}^{+\dagger}(p,s')\} = \delta_{ss'}\delta^2(\mathbf{q} - \mathbf{p})\delta(q^+ - p^+)$$
$$\{d_{TL}^{+}(q,s),\, d_{TL}^{+\dagger}(p,s')\} = \delta_{ss'}\delta^2(\mathbf{q} - \mathbf{p})\delta(q^+ - p^+)$$
$$\{b_{TL}^{+}(q,s),\, b_{TL}^{+}(p,s')\} = \{d_{TL}^{+}(q,s),\, d_{TL}^{+}(p,s')\} = 0 \qquad (10.20)$$
$$\{b_{TL}^{+\dagger}(q,s),\, b_{TL}^{+\dagger}(p,s')\} = \{d_{TL}^{+\dagger}(q,s),\, d_{TL}^{+\dagger}(p,s')\} = 0$$
$$\{b_{TL}^{+}(q,s),\, d_{TL}^{+\dagger}(p,s')\} = \{d_{TL}^{+}(q,s),\, b_{TL}^{+\dagger}(p,s')\} = 0$$
$$\{b_{TL}^{+\dagger}(q,s),\, d_{TL}^{+\dagger}(p,s')\} = \{d_{TL}^{+}(q,s),\, b_{TL}^{+}(p,s')\} = 0$$

and the spinors are

---

[56] Blaha (2006) discusses this case in detail.

$$u_{TL}^{+}(p, s) = C^- R^+ S_L(\Lambda_L(\mathbf{p}))w^1(0)$$
$$u_{TL}^{+}(p, -s) = C^- R^+ S_L(\Lambda_L(\mathbf{p}))w^2(0)$$
$$v_{TL}^{+}(p, s) = C^- R^+ S_L(\Lambda_L(\mathbf{p}))w^3(0)$$
$$v_{TL}^{+}(p, -s) = C^- R^+ S_L(\Lambda_L(\mathbf{p}))w^4(0) \tag{10.21}$$
$$u_{TL}^{++}(p, s) = w^{1T}(0)S_L^{\dagger}(\Lambda_L(\mathbf{p}))R^+C^-$$
$$u_{TL}^{++}(p, -s) = w^{2T}(0)S_L^{\dagger}(\Lambda_L(\mathbf{p}))R^+C^-$$
$$v_{TL}^{++}(p, s) = w^{3T}(0)S_L^{\dagger}(\Lambda_L(\mathbf{p}))R^+C^-$$
$$v_{TL}^{++}(p, -s) = w^{4T}(0)S_L^{\dagger}(\Lambda_L(\mathbf{p}))R^+C^-$$

where the superscript "T" indicates the transpose.

The free, "+" light-front, right-handed tachyon wave field creation/annihilation operators satisfy:

$$\{b_{TR}^{+}(q,s), b_{TR}^{++}(p,s')\} = -\delta_{ss'}\delta^2(\mathbf{q}-\mathbf{p})\delta(q^+ - p^+) \tag{10.22}$$
$$\{d_{TR}^{+}(q,s), d_{TR}^{++}(p,s')\} = -\delta_{ss'}\delta^2(\mathbf{q}-\mathbf{p})\delta(q^+ - p^+)$$
$$\{b_{TR}^{+}(q,s), b_{TR}^{+}(p,s')\} = \{d_{TR}^{+}(q,s), d_{TR}^{+}(p,s')\} = 0$$
$$\{b_{TR}^{++}(q,s), b_{TR}^{++}(p,s')\} = \{d_{TR}^{++}(q,s), d_{TR}^{++}(p,s')\} = 0$$
$$\{b_{TR}^{+}(q,s), d_{TR}^{++}(p,s')\} = \{d_{TR}^{+}(q,s), b_{TR}^{++}(p,s')\} = 0$$
$$\{b_{TR}^{++}(q,s), d_{TR}^{++}(p,s')\} = \{d_{TR}^{+}(q,s), b_{TR}^{+}(p,s')\} = 0$$

and the spinors are

$$u_{TR}^{+}(p, s) = C^+R^+u_T(p,s) \tag{10.23}$$
$$v_{TR}^{+}(p, s) = C^+R^+v_T(p,s)$$

With this selection of the quantization we can then proceed to define states.

## 10.5 Application to Duplex and Quadplex Fermion Fields

Tachyon fields are embedded in duplex and quadplex fermion field formulations described in chapters 5 – 9 of Blaha (2924j). They are one application of the light front quantization described above.

Earlier in section 9.2 we developed a UST Quadruplicated, Quadplexed Lagrangian for the Strong Interaction sector. We now have an approach that eliminates the differences between the bradyon and tachyon parts in the Lagrangian. As a result we can reexpress eq. 9.12 without the $\mathcal{H}$ matrix as:

$$\mathcal{L} = \overline{\psi}_{SBT\uparrow 2pa}{}^{\alpha\beta\rho\sigma}[M_{\uparrow a}{}^{-3}\gamma_{y\alpha\kappa}{}^{\mu}\partial/\partial y^{\mu}\ \gamma_{z\beta\lambda}{}^{\nu}\partial/\partial z^{\nu}\ \gamma_{u\rho\tau}{}^{\mu}\partial/\partial u^{\mu}\ \gamma_{v\sigma\chi}{}^{\nu}\partial/\partial v^{\nu}\ -M_{\uparrow a}]\psi_{SBT\uparrow 1pa}{}^{\kappa\lambda\tau\chi} +$$
$$+\ \overline{\psi}_{SBT\uparrow 1pa}{}^{\alpha\beta\rho\sigma}[M_{\uparrow a}{}^{-3}\gamma_{y\alpha\kappa}{}^{\mu}\partial/\partial y^{\mu}\ \gamma_{z\beta\lambda}{}^{\nu}\partial/\partial z^{\nu}\gamma_{u\rho\tau}{}^{\mu}\partial/\partial u^{\mu}\ \gamma_{v\sigma\chi}{}^{\nu}\partial/\partial v^{\nu}\ -M_{\uparrow a}]\psi_{SBT\uparrow 2pa}{}^{\kappa\lambda\tau\chi} +$$
$$+\ \overline{\psi}_{SBT\downarrow 2pa}{}^{\alpha\beta\rho\sigma}[M_{\downarrow a}{}^{-3}\gamma_{y\alpha\kappa}{}^{\mu}\partial/\partial y^{\mu}\ \gamma_{z\beta\lambda}{}^{\nu}\partial/\partial z^{\nu}\ \gamma_{u\rho\tau}{}^{\mu}\partial/\partial u^{\mu}\ \gamma_{v\sigma\chi}{}^{\nu}\partial/\partial v^{\nu}\ -M_{\downarrow a}]\psi_{SBT\downarrow 1pa}{}^{\kappa\lambda\tau\chi} +$$
$$+\ \overline{\psi}_{SBT\downarrow 1pa}{}^{\alpha\beta\rho\sigma}[M_{\downarrow a}{}^{-3}\gamma_{y\alpha\kappa}{}^{\mu}\partial/\partial y^{\mu}\ \gamma_{z\beta\lambda}{}^{\nu}\partial/\partial z^{\nu}\ \gamma_{u\rho\tau}{}^{\mu}\partial/\partial u^{\mu}\ \gamma_{v\sigma\chi}{}^{\nu}\partial/\partial v^{\nu}\ -M_{\downarrow a}]\psi_{SBT\downarrow 2pa}{}^{\kappa\lambda\tau\chi}$$
$$\tag{10.24}$$

The bradyon – tachyon distinction arises in the form of the states.

## 10.6 SBT Fermion States

The states of the quadplex field and their map to SU(4) fermions[57] was shown to have the form

**<u>Up-Type Sequence</u>**

$$e \leftrightarrow \psi_e = \psi_{1\uparrow} \sim b_y b_z t_u t_v \qquad (10.25)$$
$$u \leftrightarrow \psi_u = \psi_{2\uparrow} \sim b_y t_z t_u b_v$$
$$c \leftrightarrow \psi_c = \psi_{3\uparrow} \sim b_y b_z b_u t_v$$
$$t \leftrightarrow \psi_t = \psi_{4\uparrow} \sim b_y b_z b_u b_v$$

**<u>Down-Type Sequence</u>**

$$\nu_e \leftrightarrow \psi_e = \psi_{1\downarrow} \sim t_y t_z t_u t_v$$
$$\nu' \leftrightarrow \psi_{\nu'} = \psi_{1\downarrow} \sim b_y t_z t_u t_v$$
$$d \leftrightarrow \psi_d = \psi_{2\downarrow} \sim b_y t_z t_u b_v$$
$$s \leftrightarrow \psi_s = \psi_{3\downarrow} \sim b_y t_z b_u t_v$$
$$b \leftrightarrow \psi_b = \psi_{4\downarrow} \sim b_y t_z b_u b_v$$

where we use particle symbols to represent wave functions with the coordinate systems indicated with subscripts.

Based on the above we can identify SBT states directly. For example, the state corresponding to the u quark is

$$| \, p1,s_1; \, q2,s_2; \, q3,s_3; \, p4,s_4 > \qquad (10.26)$$

and its complex conjugate is

$$< p1,s_1; \, q2,s_2; \, q3,s_3; \, p4,s_4 | \qquad (10.27)$$

where p1 and p4 are subluminal momenta and q2 and q3 are superluminal momenta.

With this type of mapping we can take the above free Lagrangian in eq. 10.24 augmented by interactions and calculate amplitudes using a quadplex LSZ expansion in the four coordinate systems and their conjugate momentums.

*We have moved the dependence of the Lagrangian on the differentiation between bradyon and tachyon fields to its set of fermion states.*

## 10.7 A Viable SBT Formulation of the SU(4) Model

We now have simplified formulation of the UST Quadruplicated, Quadplexed Lagrangian for SU(4) using SBT, which can be directly expanded to include the ElectroWeak interactions as we showed in previous books.

---

[57] This discussion applies to SU(3)⊗U(1) with little change.

# 11. The Triumvirate of Theories

The Sequence, Cosmos and Gambol Theories combine to provide a complete, unified theory of fundamental Physical Reality. Their features interlock in many ways. Their approaches are compatible. Their consequences are decisive for the Cosmos.

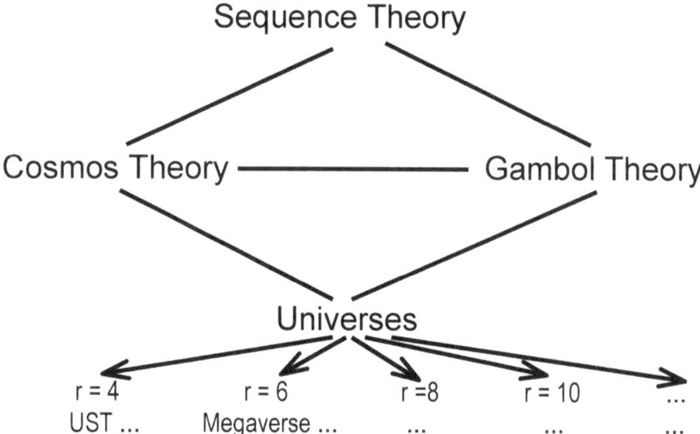


We now have a formalism for replicated fermion fields and their interactions within the UST and the Cosmos Theory framework. The complete triumvirate of theories furnishes a description that describes Physical Reality to the extent that we know it. The Consistency Conditions developed in earlier chapters accounts for the evolution of the stages of the Cosmos without a reliance on a Prime Mover.

# Appendix. Growth of the Cosmos Point to a 2–D Square[58]

## PART 1. Fractal Cosmos Curve Construction

The Cosmos Curve is constructed from a mathematical point to a unit line segment and then to a two-dimension unit square grid. The construction from a mathematical point to a line segment relies on the sum:

$$\sum_{n=1}^{\infty} 2^{-n} = 1 \qquad (A1.1)$$

The terms in this summation correspond directly to ½ of the length of columns $d_c$ in the Cosmos Theory spectrum for negative space-time dimensions. We have called this part of the Cosmos spectrum Limos.

We then construct the remainder of the fractal Cosmos Curve from line segments that order by order are of length equal ½$d_c$ for the HyperCosmos spaces in Fig. 3.1.

Thus the fully adjoined sequence of dimension array column lengths for $n = -\infty$ through $n = \infty$ builds a *Cosmos Curve from a zero dimension point to a two dimension filled square grid.*

The Cosmos Curve is a mathematical construct. We assume the column lengths corresponding to Cayley-Dickson numbers n define the rows and columns $d_{cn}$ of Cosmos space dimension arrays. Cosmos space dimension arrays are not substances. They are abstract definitions of sets of dimensions. As such they conceptually correspond to the lengths of Cosmos Curve line segments.

### A1.1 Map from Cosmos Curve Two Dimension Grid to Cosmos Theory Spaces

We now construct a relationship between a position (x, y) within the unit square of dimension two with the corresponding Cosmos space dimension array size $d_{dr}$. We propose the relation:

$$d_{dr} = 2^{r+4} = 2^{8/(x+y)} \qquad (A1.2)$$

where the coordinates x and y are confined to a two dimension unit grid by:

$$0 \le x + y \le 2 \qquad (A1.3)$$
$$0 \le x \le 1$$
$$0 \le y \le 1$$

Eq. A1.2 implies the relation between x and y and the space-time dimension r:

---

$$x + y = 8/(r + 4) \qquad\qquad (A1.4)$$

or

$$r = 8/(x + y) - 4 \qquad\qquad (A1.5)$$

We note:

$$\text{If } x + y = 2 \qquad r = 0$$

$$\text{If } x + y = 1 \qquad r = 4$$

$$\text{If } x + y = 0 \qquad r = \infty$$

### *A1.1.1 The Main Sequence of Cosmos Spaces*

If we set x = y to view the diagonal, which we call *The Main Sequence of Cosmos Spaces*, we find

$$\text{If } x = 1 \qquad r = 0$$

$$\text{If } x = \tfrac{1}{2} \qquad r = 4$$

$$\text{If } x = 0 \qquad r = \infty$$

giving a diagonal line for x and y for the entire HyperCosmos spectrum (where $r \geq 0$) with

$$r = 4/x - 4 \qquad\qquad (A1.6)$$

See Fig. A1.1. The 10 spaces of the HyperCosmos have even r values ranging from r = 0 to r = 18. The value of x corresponding to r = 18 is x = 2/11. Therefore the set of HyperCosmos spaces correspond to the x range [2/11, 1].

### *7.1.2 The Other Diagonal*

If we set x + y = 1 then we find

$$\text{If } x = 1 \text{ and } y = 0 \qquad r = 4$$

$$\text{If } x = y = \tfrac{1}{2} \qquad\qquad r = 4$$

$$\text{If } x = 0 \text{ and } y = 1 \qquad r = 4$$

giving the square grid's other diagonal line for x and y that has r = 4 for all x and y where x + y = 1. See Fig. A1.2. *This diagonal gives a certain prominence to r = 4, the dimension of our universe.*

### **A1.2 Continuity of the x = y Main Sequence Diagonal in the Unit Cosmos Grid**

The Main Sequence diagonal in the Cosmos Curve grid interpolates between the space-time dimensions r of HyperCosmos spaces. By eq. A1.6 we find

$$x = 4/(r + 4) \tag{A1.7}$$

The HyperCosmos values of r, which is even, select a set X of corresponding x values ranging from 0 ($r = \infty$) through 1 ($r = 0$). Thus we can use this square grid diagonal to obtain a continuous line of r values for spaces intermediate between HyperCosmos spaces[59] and extending to $r = \infty$. Between the spaces in the Main Sequence of Cosmos spaces are spaces that provide transitions between Cosmos Theory spaces. These spaces might play a role if travel between universes were ever to become feasible.

## A1.3 The Cosmos Grid

The points of the two dimension square grid, that we call the *Cosmos Grid*, generated by the fractal Cosmos Curve specify possible Cosmos spaces. We start with the y = x diagonal which we call the *Main Sequence of Cosmos spaces*. The Cosmos spaces points within the Cosmos Grid for the Limos spaces are described in section A2.3.

### A1.3.1 Cosmos Grid points for HyperCosmos Spaces

The Cosmos Grid points for HyperCosmos spaces are given by eqs. A1.6 and A1.5. See Fig. A1.3 for a depiction.

### A1.3.2 Cosmos Grid points for HyperCosmos Spaces of the Second Kind

The Cosmos Grid points for HyperCosmos spaces of the Second Kind have the values

$$x = 4/(r + 3) \tag{A1.8}$$

Their dimension is

$$r = 4/x - 3 \tag{A1.9}$$

The size of their associated dimension arrays is

$$d_{dr} = 2^{r+3} = 2^{4/x} \tag{A1.10}$$

### A1.3.3 Cosmos Grid Points for Cosmos Unification Spaces

The Cosmos Grid points on the Main Sequence of Cosmos spaces of the 88 Dimension UltraUnification Space and the 42 Dimension Full HyperUnification Space are

$$x = 1/23 \tag{A1.11}$$
$$x = 2/23 \tag{A1.12}$$

respectively.

### A1.3.4 Cosmos Grid Points for Cosmos Spaces

The Cosmos Grid points corresponding to Cosmos spaces are:

---

[59] Renormalization theory studies by 't Hooft, Veltman and others have used fractiona dimensions near 4.

<u>HyperCosmos Spaces</u>     where $x = 4/(r + 4)$

| $\underline{r}$ | $\underline{x}$ |
|---|---|
| 0 | 1 |
| 2 | 2/3 |
| 4 | ½ |
| 6 | 2/5 |
| 8 | 1/3 |
| ... | |
| 18 | 2/11 |
| ... | |
| $\infty$ | 0 |

<u>Second Kind HyperCosmos Spaces</u>     $x = 4/(r + 3)$

| $\underline{r}$ | $\underline{x}$ | |
|---|---|---|
| 0 | 4/3 | Off the unit grid |
| 2 | 4/5 | |
| 4 | 4/7 | |
| 6 | 4/9 | |
| 8 | 4/11 | |
| ... | | |
| 18 | 4/21 | |
| ... | | |
| $\infty$ | 0 | |

<u>Unification Spaces</u>     $x = 4/(r + 4)$

| | | |
|---|---|---|
| 42 | 2/23 | 42 Dimension Full HyperUnification Space |
| 88 | 1/23 | 88 Dimension UltraUnification Space |

Limos spaces x points are described in section A2.3.

## A1.4 Correspondence Between Cosmos Curve and Internal Symmetries

The Cosmos Curve naturally leads to HyperCosmos dimension arrays, which, in turn, lead to the dynamical Internal Symmetry interactions. We can symbolize this relation with

Cosmos Curve $\rightarrow$ Square Grid $\rightarrow$ $d_{dr}$ $\rightarrow$ dimension array $\rightarrow$ internal symmetry interactions

Correspondingly, we found the transitions:

dimension array parts $\rightarrow$ internal symmetry parts $\rightarrow$ internal symmetries $\rightarrow$ interaction terms

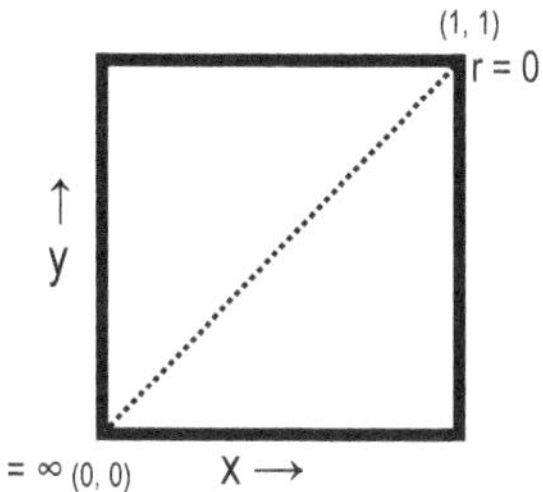


Figure A1.1. Square Cosmos Grid with a diagonal connecting r = 0 and r = ∞.

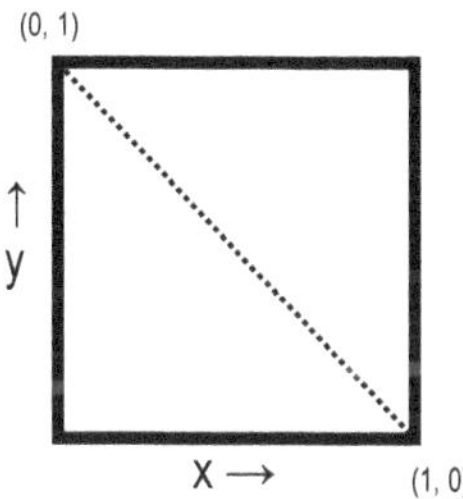


Figure A1.2. Square grid with a diagonal with each point corresponding to r = 4.  These diagonals intersect at x = y = ½ where r = 4.

# Part 2. Growing Dimensions Incrementally

This part considers growing the dimensions of a space in a continuous manner incrementally. It begins with a discussion of the Cosmic Curve's square unit grid's derivatives. Then it considers growing a dimension from r to r + 1 in increments within the framework of the grid and also within HyperCosmos spaces.

## A2.1 Derivatives of Dimensions

The continuity of the grid diagonal in the x and y variables (eqs. A1.2 – A1.6) leads us to consider derivatives of $d_{dn}$ and space-time dimension r:[60]

$$\partial d_{dn}/\partial x \quad \text{and} \quad \partial d_{dn}/\partial y \qquad (A2.1)$$

and

$$\partial r/\partial x \quad \text{and} \quad \partial r/\partial y$$

From eqs. A1.2 and A1.5 we find

$$\partial r/\partial x \; = \; \partial r/\partial y = -8/(x + y)^2 \qquad (A2.2)$$

$$\partial d_{dn}/\partial x \; \partial d_{dn}/\partial x \; = \; \partial r/\partial y \; \partial d_{dn}/\partial r \qquad (A2.3)$$

$$= -8/(x + y)^2 \; 2^{8/(x + y)}$$

and similarly for derivatives with respect to y. Note eqs. A2.2 - A2.3 imply that r increases as x + y decreases. Eq. A2.3 implies that the Cosmos' dimension array $d_{dn}$ has an essential singularity at x + y = 0. All derivatives of $d_{dn}$ with respect to x and/or y are infinite at x + y = 0.

## A2.2 Creation/Growth of Dimensions

As one traverses the x = y curve in the Cosmos Curve grid we find the dimension r continuously increases from zero to ∞ as x + y decreases to zero. It raises the question: How do the dimensions increase?

### A2.2.1 Simply Adding an Additional Dimension

One possibility is to consider the coordinate "3" dimension as growing out of a two dimensional surface. One cannot directly, infinitesimally, grow the third dimension from the two dimension surface. For an infinitesimal insertion makes the surface three dimensional regardless of being "infinitesimal."

### A2.2.2 Adding Additional Dimensions by Group Direct Products

So, in the above example, we must consider a growth path through infinitesimals that keeps the surface two-dimensional plus infinitesimal increments of a third dimension.

---

[60] The constant n is related to r by r = 2n – 2.

The spaces of the Limos sector of Cosmos Theory spaces appear to offer a solution. We note

$$d_{cn} = 2^{r/2+2} \qquad (A2.4)$$

where r is the dimension. Thus $r = 2\ln_2(d_{cn}) - 4$. We now define a new Limos space $s_i(r_{in})$ of dimension $r_{in}$ where[61]

$$r_{in} = d_{cn}/4 = 2^{r/2} \qquad (A2.5)$$

for $r < 0$. This dimension is the $r_{in}$ column in Fig. A2.1. The space $s_i(r_{in})$ may be taken to correspond to an orthogonal group $O(r_{in})$ of dimension $r_{in}$ where the orthogonal group definition is extended to fractional dimensions.[62]

Then we may take a space of dimension r and increment it with $s_i$ incrementally starting with $s_i(1/2)$ and proceeding to add to $s_i(\infty)$. The summation of the $r_{in}$ values is one by eq. 4.4 above. The product ("summation") of the infinitesimal groups totals one unit.

For example, a sum of incremental growths in dimension then may be represented as

$$SO(1, r) \otimes O(2^{-1}) \otimes O(2^{-2}) \otimes \ \ldots \otimes O(2^{-\infty}) \rightarrow SO(1, r+1) \qquad (A2.6)$$

where $O(r)$ represents the orthogonal group in a space of dimension r. The set of groups in eq. A2.6 is a subgroup of $SO(1, r+1)$ as indicated.

Thus we obtain a space with $r + 1$ spatial dimensions. While the space is incrementing, the space effectively has a fractional dimension.

The increment process adds a sequence of infinitesimal dimensions. This process may be taken to a continuous limit. Thus the $x = y$ grid diagonal may be realized as continuous in dimensions—based on an extension of the Limos definition of spaces for negative values of r.

**A2.3 Cosmos Grid Extension to Limos Spaces**

The spaces of the Limos sector of Cosmos Theory may be mapped to the Cosmos Grid using the dimension $r_{in}$ in eq. A2.5. With this definition of dimension we find

$$r_{in} = 8/(x + y) - 4 \qquad (A2.7)$$

If we assume the *Main Sequence of Cosmos spaces*: $y = x$ then

---

[61] We have previously pointed out "The dimension r determines the number of components in a dimension array. *The dimensions within a dimension array are degrees of freedom.* Dimension arrays define the spaces of Cosmos Theory. As pointed out in previous books this number r plays two roles: 1) the overall dimension of the space and 2) within a dimension array it specifies the number of space-time dimensions of the space." Thus the choice of dimension for a space is a matter of definition. For $r \geq 0$ we chose $r_{dimension\ array} = r$. We chose $r_{dimension\ array} = r_{in}$ for negative dimensions r.

[62] We have considered groups with fractional dimensions in past books for the case of the unitary groups.

$$r_{in} = 4/x - 4 = 2^{r/2} \qquad (A2.8)$$

where r is the space-time dimension column in Fig. A2.1 and

$$x = 4/(r_{in} + 4) = 4/(2^{r/2} + 4) \qquad (A2.9)$$

Some x, $r_{in}$ and r values for Limos spaces

$$(A2.10)$$

| r | x | $r_{in}$ |
|---|---|---|
| -2 | 8/9 | ½ |
| -4 | 16/17 | 1/4 |
| ... | | |
| $-\infty$ | 1 | 0 |

Thus the Limos spaces have x values ranging from 8/9 to 1 on the Cosmos Grid Main Sequence.

| Blaha Space Number $N = o_s$ | Cayley-Dickson Number $n$ | Cayley Number $d_c$ | Dimension Array column length $d_{cn}$ | Dimension Array Size $d_{dn}$ | Space-time-Dimension $r$ | Incremental Dimension $r_{in}$ |
|---|---|---|---|---|---|---|
| Limos : | | | | | | |
| 10 | 0 | 1 | 2 | $2^2$ | -2 | ½ |
| 11 | -1 | ½ | 1 | 1 | -4 | ¼ |
| 12 | -2 | ¼ | ½ | $½^2$ | -6 | 1/8 |
| 13 | -3 | 1/8 | ¼ | $¼^2$ | -8 | 1/16 |
| 14 | -4 | 1/16 | 1/8 | $1/8^2$ | -10 | 1/32 |
| ⋮ | | | | | | |

Figure A2.1 Limos Sector of the Cosmos Theory set of spaces from Fig. 3.1. Note the additional $r_{in}$ column.

## A2.4 Growth of the Cosmos Curve

We now consider growing the Cosmos Curve order by order incrementally. The orders of piecewise linear line segment lengths in the construction of a Cosmos Curve are related to the Hilbert curve construction:[63]

Cosmos Cayley-Dickson number $n = n_H - 1$
Hilbert Curve Line Length $= 2^{n_H}$
Hilbert Number of "boxes" $= 2^{2n_H}$
Cosmos Dimension Array Column Length $= 2^{n+1} = 2^{n_H}$
Cosmos Dimension Array Size (number of elements in array) $= 2^{2n+2} = 2^{2n_H}$

---

[63] The construction of the fractal curve corresponding to Cosmos Theory spaces was shown to be similar to the construction of the Hilbert fractal curve in Blaha (2024b).

where n is the Cosmos Cayley-Dickson number and $n_H$ is the order in the Hilbert curve construction.

We can create an infinitesimal growth sequence from order to order by infinitesimal increments:

$$2^{n+1} \rightarrow 2^{n+2}$$

where the increments are

$$2^{n+1-k}$$

for $k = \infty$ through 1 based on the identity

$$2^{n+1} \sum_{k=1}^{\infty} 2^{-k} = 2^{n+1} \qquad (A2.11)$$

The incremental growth in piecewise linear line segment lengths follows from order to order of the Cosmos Curve.


# Part 3. Generation of Cosmos Theory  Spaces from the Cosmos Grid

The HyperCosmos spaces of Cosmos Theory can be generated from the fractal Cosmos Curve. Thereby, we obtain a full specification of dimensions of the HyperCosmos Spaces starting from a zero dimension point.

Our goal was to start with the simplest possible origin—a zero dimension point that a Creator might choose and then proceed to create the Cosmos. The search for simplicity dates back to Pre-Socratic times.

## A3.1 Dimensions of Stages in the Construction of the Cosmos Curve

We start with a zero dimension point and then generate a one dimension unit line segment. Then we follow a fractal construction process to create a two dimension unit square grid.

The unit square grid encompasses all the HyperCosmos spaces within it.

## A3.2 Extraction of Dimension Arrays of HyperCosmos Spaces

We then use the fractal grid decomposition process to extract the dimension array size $d_{nr}$ for all HyperCosmos spaces. Thus we have the almost complete set of Cosmos spaces.

## A3.3 HyperCosmos Spaces of the Second Kind

The remaining spaces of Cosmos Theory, Limos spaces and HyperCosmos spaces of the Second Kind, may be generated in the unification process. See section A1.3.2 for the Cosmos Grid points of HyperCosmos spaces of the Second Kind.

## A3.4 Limos Sector Spaces

See section A2.3.

## A3.5 Cosmos Unification Spaces

See section A1.3.3 for the Cosmos Grid points of Cosmos Unification spaces.

## A3.4 Universes

The Cosmos Theory structure built in the above outlined manner may then be used in the construction of universes either by a creator or dynamically as discussed earlier in this book.

*Thus we have generated a Cosmos from a Mathematical point.*

# REFERENCES


Akhiezer, N. I., Frink, A. H. (tr), 1962, *The Calculus of Variations* (Blaisdell Publishing, New York, 1962).

Bjorken, J. D., Drell, S. D., 1964, *Relativistic Quantum Mechanics* (McGraw-Hill, New York, 1965).

Bjorken, J. D., Drell, S. D., 1965, *Relativistic Quantum Fields* (McGraw-Hill, New York, 1965).

Blaha, S., 1995, *C++ for Professional Programming* (International Thomson Publishing, Boston, 1995).

_______, 1998, *Cosmos and Consciousness* (Pingree-Hill Publishing, Auburn, NH, 1998 and 2002).

_______, 2002, *A Finite Unified Quantum Field Theory of the Elementary Particle Standard Model and Quantum Gravity Based on New Quantum Dimensions™ & a New Paradigm in the Calculus of Variations* (Pingree-Hill Publishing, Auburn, NH, 2002).

_______, 2004, *Quantum Big Bang Cosmology: Complex Space-time General Relativity, Quantum Coordinates™ Dodecahedral Universe, Inflation, and New Spin 0, ½, 1 & 2 Tachyons & Imagyons* (Pingree-Hill Publishing, Auburn, NH, 2004).

_______, 2005a, *Quantum Theory of the Third Kind: A New Type of Divergence-free Quantum Field Theory Supporting a Unified Standard Model of Elementary Particles and Quantum Gravity based on a New Method in the Calculus of Variations* (Pingree-Hill Publishing, Auburn, NH, 2005).

_______, 2005b, *The Metatheory of Physics Theories, and the Theory of Everything as a Quantum Computer Language* (Pingree-Hill Publishing, Auburn, NH, 2005).

_______, 2005c, *The Equivalence of Elementary Particle Theories and Computer Languages: Quantum Computers, Turing Machines, Standard Model, Superstring Theory, and a Proof that Gödel's Theorem Implies Nature Must Be Quantum* (Pingree-Hill Publishing, Auburn, NH, 2005).

_______, 2006a, *The Foundation of the Forces of Nature* (Pingree-Hill Publishing, Auburn, NH, 2006).

_______, 2006b, *A Derivation of ElectroWeak Theory based on an Extension of Special Relativity; Black Hole Tachyons; & Tachyons of Any Spin.* (Pingree-Hill Publishing, Auburn, NH, 2006).

_______, 2007a, *Physics Beyond the Light Barrier: The Source of Parity Violation, Tachyons, and A Derivation of Standard Model Features* (Pingree-Hill Publishing, Auburn, NH, 2007).

_______, 2007b, *The Origin of the Standard Model: The Genesis of Four Quark and Lepton Species, Parity Violation, the ElectroWeak Sector, Color SU(3), Three Visible Generations of Fermions, and One Generation of Dark Matter with Dark Energy* (Pingree-Hill Publishing, Auburn, NH, 2007).

_______, 2008a, *A Direct Derivation of the Form of the Standard Model From GL(16)* (Pingree-Hill Publishing, Auburn, NH, 2008).

_______, 2008b, *A Complete Derivation of the Form of the Standard Model With a New Method to Generate Particle Masses Second Edition* (Pingree-Hill Publishing, Auburn, NH, 2008)

_______, 2009, *The Algebra of Thought & Reality: The Mathematical Basis for Plato's Theory of Ideas, and Reality Extended to Include A Priori Observers and Space-Time Second Edition* (Pingree-Hill Publishing, Auburn, NH, 2009).

_______, 2009a, *Bright Stars, Bright Universe* (Pingree-Hill Publishing, Auburn, NH, 2008)

_______, 2010a, *Operator Metaphysics: A New Metaphysics Based on a New Operator Logic and a New Quantum Operator Logic that Lead to a Mathematical Basis for Plato's Theory of Ideas and Reality* (Pingree-Hill Publishing, Auburn, NH, 2010).

_______, 2010b, *The Standard Model's Form Derived from Operator Logic, Superluminal Transformations and GL(16)* (Pingree-Hill Publishing, Auburn, NH, 2010).

_______, 2010c, *SuperCivilizations: Civilizations as Superorganisms* (McMann-Fisher Publishing, Auburn, NH, 2010).

_______, 2011a, *21st Century Natural Philosophy Of Ultimate Physical Reality* (McMann-Fisher Publishing, Auburn, NH, 2011).

_______, 2011b, *All the Universe! Faster Than Light Tachyon Quark Starships & Particle Accelerators with the LHC as a Prototype Starship Drive Scientific Edition* (Pingree-Hill Publishing, Auburn, NH, 2011).

_______, 2011c, *From Asynchronous Logic to The Standard Model to Superflight to the Stars* (Blaha Research, Auburn, NH, 2011).

_______, 2012a, *From Asynchronous Logic to The Standard Model to Superflight to the Stars volume 2: Superluminal CP and CPT, U(4) Complex General Relativity and The Standard Model, Complex Vierbein General Relativity, Kinetic Theory, Thermodynamics* (Blaha Research, Auburn, NH, 2012).

______, 2012b, *Standard Model Symmetries, And Four And Sixteen Dimension Complex Relativity; The Origin Of Higgs Mass Terms* (Blaha Reasearch, Auburn, NH, 2012).

______, 2013a, *Multi-Stage Space Guns, Micro-Pulse Nuclear Rockets, and Faster-Than-Light Quark-Gluon Ion Drive Starships* (Blaha Research, Auburn, NH, 2013).

______, 2013b, *The Bridge to Dark Matter; A New Sibling Universe; Dark Energy; Inflatons; Quantum Big Bang; Superluminal Physics; An Extended Standard Model Based on Geometry* (Blaha Reasearch, Auburn, NH, 2013).

______, 2014a, *Universes and Megaverses: From a New Standard Model to a Physical Megaverse; The Big Bang; Our Sibling Universe's Wormhole; Origin of the Cosmological Constant, Spatial Asymmetry of the Universe, and its Web of Galaxies; A Baryonic Field between Universes and Particles; Megaverse Extended Wheeler-DeWitt Equation* (Blaha Reasearch, Auburn, NH, 2014).

______, 2014b, *All the Megaverse! Starships Exploring the Endless Universes of the Cosmos Using the Baryonic Force* (Blaha Research, Auburn, NH, 2014).

______, 2014c, *All the Megaverse! II Between Megaverse Universes: Quantum Entanglement Explained by the Megaverse Coherent Baryonic Radiation Devices – PHASERs Neutron Star Megaverse Slingshot Dynamics Spiritual and UFO Events, and the Megaverse Microscopic Entry into the Megaverse* (Blaha Research, Auburn, NH, 2014).

______, 2015a, *PHYSICS IS LOGIC PAINTED ON THE VOID: Origin of Bare Masses and The Standard Model in Logic, U(4) Origin of the Generations, Normal and Dark Baryonic Forces, Dark Matter, Dark Energy, The Big Bang, Complex General Relativity, A Megaverse of Universe Particles* (Blaha Research, Auburn, NH, 2015).

______, 2015b, *PHYSICS IS LOGIC Part II: The Theory of Everything, The Megaverse Theory of Everything, U(4)⊗U(4) Grand Unified Theory (GUT), Inertial Mass = Gravitational Mass, Unified Extended Standard Model and a New Complex General Relativity with Higgs Particles, Generation Group Higgs Particles* (Blaha Research, Auburn, NH, 2015).

______, 2015c, *The Origin of Higgs ("God") Particles and the Higgs Mechanism: Physics is Logic III, Beyond Higgs – A Revamped Theory With a Local Arrow of Time, The Theory of Everything Enhanced, Why Inertial Frames are Special, Universes of the Mind* (Blaha Research, Auburn, NH, 2015).

______, 2015d, *The Origin of the Eight Coupling Constants of The Theory of Everything: U(8) Grand Unified Theory of Everything (GUTE), $S^8$ Coupling Constant Symmetry, Space-Time Dependent Coupling Constants, Big Bang Vacuum Coupling Constants, Physics is Logic IV* (Blaha Research, Auburn, NH, 2015).

______, 2016a, *New Types of Dark Matter, Big Bang Equipartition, and A New U(4) Symmetry in the Theory of Everything: Equipartition Principle for Fermions, Matter is 83.33% Dark, Penetrating the Veil of the Big Bang, Explicit QFT Quark Confinement and Charmonium, Physics is Logic V* (Blaha Research, Auburn, NH, 2016).

______, 2016b, *The Periodic Table of the 192 Quarks and Leptons in The Theory of Everything: The U(4) Layer Group, Physics is Logic VI* (Blaha Research, Auburn, NH, 2016).

______, 2016c, *New Boson Quantum Field Theory, Dark Matter Dynamics, Dark Matter Fermion Layer Mixing, Genesis of Higgs Particles, New Layer Higgs Masses, Higgs Coupling Constants, Non-Abelian Higgs Gauge Fields, Physics is Logic VII* (Blaha Research, Auburn, NH, 2016).

______, 2016d, *Unification of the Strong Interactions and Gravitation: Quark Confinement Linked to Modified Short-Distance Gravity; Physics is Logic VIII* (Blaha Research, Auburn, NH, 2016).

______, 2016e, *MoND: Unification of the Strong Interactions and Gravitation II, Quark Confinement Linked to Large-Scale Gravity, Physics is Logic IX* (Blaha Research, Auburn, NH, 2016).

______, 2016f, *CQ Mechanics: A Unification of Quantum & Classical Mechanics, Quantum/Semi-Classical Entanglement, Quantum/Classical Path Integrals, Quantum/Classical Chaos* (Blaha Research, Auburn, NH, 2016).

______, 2016g, *GEMS Unified Gravity, ElectroMagnetic and Strong Interactions: Manifest Quark Confinement, A Solution for the Proton Spin Puzzle, Modified Gravity on the Galactic Scale* (Pingree Hill Publishing, Auburn, NH, 2016).

______, 2016h, *Unification of the Seven Boson Interactions based on the Riemann-Christoffel Curvature Tensor* (Pingree Hill Publishing, Auburn, NH, 2016).

______, 2017a, *Unification of the Eleven Boson Interactions based on 'Rotations of Interactions'* (Pingree Hill Publishing, Auburn, NH, 2017).

______, 2017b, *The Origin of Fermions and Bosons, and Their Unification* (Pingree Hill Publishing, Auburn, NH, 2017).

______, 2017c, *Megaverse: The Universe of Universes* (Pingree Hill Publishing, Auburn, NH, 2017).

______, 2017d, *SuperSymmetry and the Unified SuperStandard Model* (Pingree Hill Publishing, Auburn, NH, 2017).

______, 2017e, *From Qubits to the Unified SuperStandard Model with Embedded SuperStrings: A Derivation* (Pingree Hill Publishing, Auburn, NH, 2017).

______, 2017f, *The Unified SuperStandard Model in Our Universe and the Megaverse: Quarks, … ,* (Pingree Hill Publishing, Auburn, NH, 2017).

______, 2018a, *The Unified SuperStandard Model and the Megaverse SECOND EDITION A Deeper Theory based on a New Particle Functional Space that Explicates Quantum Entanglement Spookiness (Volume 1)* (Pingree Hill Publishing, Auburn, NH, 2018).

______, 2018b, *Cosmos Creation: The Unified SuperStandard Model, Volume 2, SECOND EDITION* (Pingree Hill Publishing, Auburn, NH, 2018).

______, 2018c, *God Theory* (Pingree Hill Publishing, Auburn, NH, 2018).

______, 2018d, *Immortal Eye: God Theory: Second Edition* (Pingree Hill Publishing, Auburn, NH, 2018).

______, 2018e, *Unification of God Theory and Unified SuperStandard Model THIRD EDITION* (Pingree Hill Publishing, Auburn, NH, 2018).

______, 2019a, *Calculation of: QED α = 1/137, and Other Coupling Constants of the Unified SuperStandard Theory* (Pingree Hill Publishing, Auburn, NH, 2019).

______, 2019b, *Coupling Constants of the Unified SuperStandard Theory SECOND EDITION* (Pingree Hill Publishing, Auburn, NH, 2019).

______, 2019c, *New Hybrid Quantum Big_Bang–Megaverse_Driven Universe with a Finite Big Bang and an Increasing Hubble Constant* (Pingree Hill Publishing, Auburn, NH, 2019).

______, 2019d, *The Universe, The Electron and The Vacuum* (Pingree Hill Publishing, Auburn, NH, 2019).

______, 2019e, *Quantum Big Bang – Quantum Vacuum Universes (Particles)* (Pingree Hill Publishing, Auburn, NH, 2019).

______, 2019f, *The Exact QED Calculation of the Fine Structure Constant Implies ALL 4D Universes have the Same Physics/Life Prospects* (Pingree Hill Publishing, Auburn, NH, 2019).

______, 2019g, *Unified SuperStandard Theory and the SuperUniverse Model: The Foundation of Science* (Pingree Hill Publishing, Auburn, NH, 2019).

______, 2020a, *Quaternion Unified SuperStandard Theory (The QUeST) and Megaverse Octonion SuperStandard Theory (MOST)* (Pingree Hill Publishing, Auburn, NH, 2020).

______, 2020b, *United Universes Quaternion Universe - Octonion Megaverse* (Pingree Hill Publishing, Auburn, NH, 2020).

______, 2020c, *Unified SuperStandard Theories for Quaternion Universes & The Octonion Megaverse* (Pingree Hill Publishing, Auburn, NH, 2020).

______, 2020d, *The Essence of Eternity: Quaternion & Octonion SuperStandard Theories* (Pingree Hill Publishing, Auburn, NH, 2020).

______, 2020e, *The Essence of Eternity II* (Pingree Hill Publishing, Auburn, NH, 2020).

______, 2020f, *A Very Conscious Universe* (Pingree Hill Publishing, Auburn, NH, 2020).

______, 2020g, *Hypercomplex Universe* (Pingree Hill Publishing, Auburn, NH, 2020).

______, 2020h, *Beneath the Quaternion Universe* (Pingree Hill Publishing, Auburn, NH, 2020).

______, 2020i, *Why is the Universe Real? From Quaternion & Octonion to Real Coordinates* (Pingree Hill Publishing, Auburn, NH, 2020).

______, 2020j, *The Origin of Universes: of Quaternion Unified SuperStandard Theory (QUeST); and of the Octonion Megaverse (UTMOST)* (Pingree Hill Publishing, Auburn, NH, 2020).

______, 2020k, *The Seven Spaces of Creation: Octonion Cosmology* (Pingree Hill Publishing, Auburn, NH, 2020).

______, 2020l, *From Octonion Cosmology to the Unified SuperStandard Theory of Particles* (Pingree Hill Publishing, Auburn, NH, 2020).

______, 2021a, *Pioneering the Cosmos* (Pingree Hill Publishing, Auburn, NH, 2021).

______, 2021b, *Pioneering the Cosmos II* (Pingree Hill Publishing, Auburn, NH, 2021).

______, 2021c, *Beyond Octonion Cosmology* (Pingree Hill Publishing, Auburn, NH, 2021).

______, 2021d, *Universes are Particles* (Pingree Hill Publishing, Auburn, NH, 2021).

______, 2021e, *Octonion-like dna-based life, Universe expansion is decay, Emerging New Physics* (Pingree Hill Publishing, Auburn, NH, 2021).

______, 2021f, *The Science of Creation New Quantum Field Theory of Spaces* (Pingree Hill Publishing, Auburn, NH, 2021).

______, 2021g, *Quantum Space Theory With Application to Octonion Cosmology & Possibly To Fermionic Condensed Matter* (Pingree Hill Publishing, Auburn, NH, 2021).

______, 2021h, *21ˢᵗ Century Natural Philosophy of Octonion Cosmology , and Predestination, Fate, and Free Will* (Pingree Hill Publishing, Auburn, NH, 2021).

______, 2021i, *Beyond Octonion Cosmology II : Origin of the Quantum; A New Generalized Field Theory (GiFT); A Proof of the Spectrum of Universes; Atoms in Higher Universes* (Pingree Hill Publishing, Auburn, NH, 2021).

______, 2021j, *Integration of General Relativity and Quantum Theory: Octonion Cosmology, GiFT, Creation/Annihilation Spaces CASe, Reduction of Spaces to a Few Fermions and Symmetries in Fundamental Frames* (Pingree Hill Publishing, Auburn, NH, 2021).

______, 2022a, *New View of Octonion Cosmology Based on the Unification of General Relativity and Quantum Theory* (Pingree Hill Publishing, Auburn, NH, 2022).

______, 2022b, *The  Dust Beneath Hypercomplex Cosmology* (Pingree Hill Publishing, Auburn, NH, 2022).

______, 2022c, *Passing Through Nature to Eternity: ProtoCosmos, HyperCosmos, Unified SuperStandard Theory* (Pingree Hill Publishing, Auburn, NH, 2022).

______, 2022d, *HyperCosmos Fractionation and Fundamental Reference Frame Based Unification: Particle Inner Space Basis of Parton and Dual Resonance Models* (Pingree Hill Publishing, Auburn, NH, 2022).

______, 2022e, *A New UniDimension ProtoCosmos and SuperString F-Theory Relation to the HyperCosmos* (Pingree Hill Publishing, Auburn, NH, 2022).

______, 2022f, *The Cosmic Panorama: ProtoCosmos, HyperCosmos,Unified SuperStandard Theory (UST) Derivation* (Pingree Hill Publishing, Auburn, NH, 2022).

______, 2022g, *Ultimate Origin: ProtoCosmos and HyperCosmos* (Pingree Hill Publishing, Auburn, NH, 2022).

______, 2023a, *UltraUnification and the Generation of the Cosmos* (Pingree Hill Publishing, Auburn, NH, 2023).

______, 2023b, *God and and Cosmos Theory* (Pingree Hill Publishing, Auburn, NH, 2023).

______, 2023c, *A New Completely Geometric SU(8) Cosmos Theory; New PseudoFermion Fields; Fibonacci-like Dimension Arrays; Ramsey Number Approximation* (Pingree Hill Publishing, Auburn, NH, 2023).

______, 2023d, *Newton's Apple is Now the Fermion* (Pingree Hill Publishing, Auburn, NH, 2023).

______, 2023e,*Cosmos Theory: The Sub-Particle Gambol Model* (Pingree Hill Publishing, Auburn, NH, 2023).

______, 2024a, *Cosmos-Universe-Particle-Gambol Theory* (Pingree Hill Publishing, Auburn, NH, 2024).

______, 2024b, *Fractal Cosmos Theory* (Pingree Hill Publishing, Auburn, NH, 2024).

______, 2024c, *Fractal Cosmic Curve: Tensor-Based CosmosTheory* (Pingree Hill Publishing, Auburn, NH, 2024).

______, 2024d, *The Eternal Form of Cosmos Theory* (Pingree Hill Publishing, Auburn, NH, 2024).

______, 2024e, *The Eternal Form of Cosmos Theory Third Edition* (Pingree Hill Publishing, Auburn, NH, 2024).

______, 2024f, *Fundamental Constants of Cosmos Theory and The Standard Model* (Pingree Hill Publishing, Auburn, NH, 2024).

______, 2024g, *Quark, Lepton, W and Z Masses of Cosmos Theory and The Standard Model* (Pingree Hill Publishing, Auburn, NH, 2024).

______, 2024h, *Geometric Cosmos Geometric Universe* (Pingree Hill Publishing, Auburn, NH, 2024).

______, 2024i, *Particles and Universes of Cosmos Theory* (Pingree Hill Publishing, Auburn, NH, 2024).

______, 2024j, *Unification of the Subluminal and the Superluminal in Cosmos Theory* (Pingree Hill Publishing, Auburn, NH, 2024).

______, 2024k, *The Dawn of Dynamic Cosmos Dimension Arrays* (Pingree Hill Publishing, Auburn, NH, 2024).

______, 2024ℓ, *Structure and Dynamics of Cosmos Theory and the Unified SuperStandard Theory* (Pingree Hill Publishing, Auburn, NH, 2024).

______, 2025a, *Black Holes, White Holes, and Superluminal Starship* (Pingree Hill Publishing, Auburn, NH, 2025).

______, 2025b, *The Sequence Theory Foundation of Cosmos Theory and Gambol Theory* (Pingree Hill Publishing, Auburn, NH, 2025).

Eddington, A. S., 1952, *The Mathematical Theory of Relativity* (Cambridge University Press, Cambridge, U.K., 1952).

Fant, Karl M., 2005, *Logically Determined Design: Clockless System Design With NULL Convention Logic* (John Wiley and Sons, Hoboken, NJ, 2005).

Feinberg, G. and Shapiro, R., 1980, *Life Beyond Earth: The Intelligent Earthlings Guide to Life in the Universe* (William Morrow and Company, New York, 1980).

Gelfand, I. M., Fomin, S. V., Silverman, R. A. (tr), 2000, *Calculus of Variations* (Dover Publications, Mineola, NY, 2000).

Giaquinta, M., Modica, G., Souchek, J., 1998, *Cartesian Coordinates in the Calculus of Variations* Volumes I and II (Springer-Verlag, New York, 1998).

Giaquinta, M., Hildebrandt, S., 1996, *Calculus of Variations* Volumes I and II (Springer-Verlag, New York, 1996).

Gradshteyn, I. S. and Ryzhik, I. M., 1965, *Table of Integrals, Series, and Products* (Academic Press, New York, 1965).

Heitler, W., 1954, *The Quantum Theory of Radiation* (Claendon Press, Oxford, UK, 1954).

Huang, Kerson, 1992, *Quarks, Leptons & Gauge Fields 2$^{nd}$ Edition* (World Scientific Publishing Company, Singapore, 1992).

Jost, J., Li-Jost, X., 1998, *Calculus of Variations* (Cambridge University Press, New York, 1998).

Kaku, Michio, 1993, *Quantum Field Theory*, (Oxford University Press, New York, 1993).

Kirk, G. S. and Raven, J. E., 1962, *The Presocratic Philosophers* (Cambridge University Press, New York, 1962).

Landau, L. D. and Lifshitz, E. M., 1987, *Fluid Mechanics 2$^{nd}$ Edition*, (Pergamon Press, Elmsford, NY, 1987).

Rescher, N., 1967, *The Philosophy of Leibniz* (Prentice-Hall, Englewood Cliffs, NJ, 1967).

Riesz, Frigyes and Sz.-Nagy, Béla, 1990, *Functional Analysis* ( Dover Publications, New York, 1990).

Sakurai, J. J., 1964, *Invariance Principles and Elementary Particles* (Princeton University Press, Princeton, NJ, 1964).

Weinberg, S., 1972, *Gravitation and Cosmology* (John Wiley and Sons, New York, 1972).

Weinberg, S., 1995, *The Quantum Theory of Fields Volume I* (Cambridge University Press, New York, 1995).

# INDEX

Stephen Blaha is a well-known Physicist and Man of Letters with interests in Science, Society and civilization, the Arts, and Technology. He had an Alfred P. Sloan Foundation scholarship in college. He received his Ph.D. in Physics from Rockefeller University. He has served on the faculties of several major universities. He was also a Member of the Technical Staff at Bell Laboratories, a manager at the Boston Globe Newspaper, a Director at Wang Laboratories, and President of Blaha Software Inc. and of Janus Associates Inc. (NH).

Among other achievements he was a co-discoverer of the "r potential" for heavy quark binding developing the first (and still the only demonstrable) non-Aeolian gauge theory with an "r" potential; first suggested the existence of topological structures in superfluid He-3; first proposed Yang-Mills theories would appear in condensed matter phenomena with non-scalar order parameters; first developed a grammar-based formalism for quantum computers and applied it to elementary particle theories; first developed a new form of quantum field theory without divergences (thus solving a major 60 year old problem that enabled a unified theory of the Standard Model and Quantum Gravity without divergences to be developed); first developed a formulation of complex General Relativity based on analytic continuation from real space-time; first developed a generalized non-homogeneous Robertson-Walker metric that enabled a quantum theory of the Big Bang to be developed without singularities at t = 0; first generalized Cauchy's theorem and Gauss' theorem to complex, curved multi-dimensional spaces; received Honorable Mention in the Gravity Research Foundation Essay Competition in 1978; first developed a physically acceptable theory of faster-than-light particles; first derived a composition of extremums method in the Calculus of Variations; first quantitatively suggested that inflationary periods in the history of the universe were not needed; first proved Gödel's Theorem implies Nature must be quantum; provided a new alternative to the Higgs Mechanism, and Higgs particles, to generate masses; first showed how to resolve logical paradoxes including Gödel's Undecidability Theorem by developing Operator Logic and Quantum Operator Logic; first developed a quantitative harmonic oscillator-like model of the life cycle, and interactions, of civilizations; first showed how equations describing superorganisms also apply to civilizations. A recent book shows his theory applies successfully to the past 14 years of history and to *new* archaeological data on Andean and Mayan civilizations as well as Early Anatolian and Egyptian civilizations.

He first developed an axiomatic derivation of the form of The Standard Model from geometry – space-time properties – The Unified SuperStandard Model. It unifies all the known forces of Nature. It also has a Dark Matter sector that includes a Dark ElectroWeak sector with Dark doublets and Dark gauge interactions. It uses quantum coordinates to remove infinities that crop up in most interacting quantum field theories and additionally to remove the infinities that appear in the Big Bang and generate inflationary growth of the universe. It shows gravity has a MOND-like form without sacrificing Newton's Laws. It relates the interactions of the MOND-like sector of gravity with the r-potential of Quark Confinement. The axioms of the theory lead to the question of their origin. We suggest in the preceding edition of this book it can be attributed to an entity with God-like properties. We explore these properties in "God Theory" and show they predict that the Cosmos exists forever although individual universes (or incarnations of our universe) "come and go." Several other important results emerge from God Theory such a functionally triune God. The Unified SuperStandard Theory has many other important parts described in the Current Edition of *The Unified SuperStandard Theory* and expanded in subsequent volumes.

Blaha has had a major impact on a succession of elementary particle theories: his Ph.D. thesis (1970), and papers, showed that quantum field theory calculations to all orders in ladder approximations could not give scaling deep inelastic electron-nucleon scattering. He later showed the eigenvalue equation for the fine structure constant $\alpha$ in Johnson-Baker-Willey QED had a zero at $\alpha = 1$ not 1/137 by solving the Schwinger-Dyson equations to all orders in an approximation that agreed with exact results to 4$^{th}$ order in $\alpha$ thus ending interest in this theory. In 1979 at Prof. Ken Johnson's (MIT) suggestion he calculated the proton-neutron mass difference in the MIT bag model and found the result had the wrong sign reducing interest in the bag model. These results all appear in Physical Review papers. In the 2000's he repeatedly pointed out the shortcomings of SuperString theory and showed that The Standard Model's form could be derived from space-time geometry by an extension of Lorentz transformations to faster than light transformations. This deeper space-time basis greatly increases the possibility that it is part of THE fundamental theory. Recently, Blaha showed that the Weak interactions differed significantly from the Strong, electromagnetic and gravitation interactions in important respects while these interactions had similar features, and suggested that ElectroWeak theory, which is essentially a glued union of the Weak interactions and Electromagnetism, possibly modulo unknown Higgs particle features, be replaced by a unified theory of the other interactions combined with a stand-alone Weak interaction theory. Blaha also showed that, if Charmonium calculations are taken seriously, the Strong interaction coupling constant is only a factor of five larger than the electromagnetic coupling constant, and thus Strong interaction perturbation theory would make sense and yield physically meaningful results.

In graduate school (1965-71) he wrote substantial papers in elementary particles and group theory: The Inelastic E- P Structure Functions in a Gluon Model. Phys. Lett. B40:501-502,1972; Deep-Inelastic E-P Structure Functions In A Ladder Model With Spin 1/2 Nucleons, Phys.Rev. D3:510-523,1971; Continuum Contributions To The Pion Radius, Phys. Rev. 178:2167-2169,1969; Character Analysis of U(N) and SU(N), J. Math. Phys. <u>10</u>, 2156 (1969); and The Calculation of the Irreducible Characters of the Symmetric Group in Terms of the Compound Characters, (Published as Blaha's Lemma in D. E. Knuth's book: *The Art of Computer Programming Vols. 1 – 4*).

In the early 1980's Blaha was also a pioneer in the development of UNIX for financial, scientific and Internet applications: benchmarked UNIX versions showing that block size was critical for UNIX performance, developing financial modeling software, starting database benchmarking comparison studies, developing Internet-like UNIX networking (1982) and developing a hybrid shell programming technique (1982) that was a precursor to the PERL programming language. He was also the manager of the AT&T ten-year future products development database. His work helped lead to commercial UNIX on computers such as Sun Micros, IBM AIX minis, and Apple computers.

In the 1980's he pioneered the development of PC Desktop Publishing on laser printers and was nominated for three "Awards for Technical Excellence" in 1987 by PC Magazine for PC software products that he designed and developed.

Recently he has developed a theory of Megaverses – actual universes of which our universe is one – with quantum particle-like properties based on the Wheeler-DeWitt equation of Quantum Gravity. He has developed a theory of a baryonic force, which had been conjectured many years ago, and estimated the strength of the force based on discrepancies in measurements of the gravitational constant G. This force, operative in D-dimensional space, can be used to escape from our universe in "uniships" which are the equivalent of the faster-than-light starships proposed in the author's earlier books. Thus travel to other universes, as well as to other stars is possible.

Blaha also considered the complexified Wheeler-DeWitt equation and showed that its limitation to real-valued coordinates and metrics generated a Cosmological Constant in the Einstein equations.

The author has also recently written a series of books on the serious problems of the United States and their solution as well as a book on the decline of Mankind that will follow from current social and genetic trends in Mankind.

In the past twenty years Dr. Blaha has written over 80 books on a wide range of topics. Some recent major works are: *From Asynchronous Logic to The Standard Model to Superflight to the Stars, All the Universe!, SuperCivilizations: Civilizations as Superorganisms, America's Future: an Islamic Surge, ISIS, al Qaeda, World Epidemics, Ukraine, Russia-China Pact, US Leadership Crisis, The Rises and Falls of Man – Destiny – 3000 AD: New Support for a Superorganism MACRO-THEORY of CIVILIZATIONS From CURRENT WORLD TRENDS and NEW Peruvian, Pre-Mayan, Mayan, Anatolian, and Early Egyptian Data, with a Projection to 3000 AD,* and *Mankind in Decline: Genetic Disasters, Human-Animal Hybrids, Overpopulation, Pollution, Global Warming, Food and Water Shortages, Desertification, Poverty, Rising Violence, Genocide, Epidemics, Wars, Leadership Failure.*

He has taught approximately 4,000 students in undergraduate, graduate, and postgraduate corporate education courses primarily in major universities, and large companies and government agencies.

He developed a quantum theory, The Unified SuperStandard Theory (UST), which describes elementary particles in detail without the difficulties of conventional quantum field theory. He found that the internal symmetries of this theory could be exactly derived from an octonion theory called QUeST. He further found that another octonion theory (UTMOST) describes the Megaverse. It can hold QUeST universes such as our own universe. It has an internal symmetry structure which is a superset of the QUeST internal symmetries.

Recently he developed Octonion Cosmology. He replaced it with HyperCosmos theory, which has significantly better features. He developed a fractionalization process for dimensions, particles and symmetry groups. He also described transformation that reduced particles and dimensions to a far more compact form. He also developed a precursor theory ProtoCosmos that leads to the HyperCosmos.

The author showed that space-time and Internal Symmetries can be unified in any of the ten HyperCosmos spaces in their associated HyperUnification spaces. The combined set of HyperUnification spaces enable all HyperCosmos dimensions to be obtained by a General Relativistic transformation from one primordial dimension in the 42 space-time dimension unified HyperUnification space.

At present the author devel;oped the Cosmos Theory that incorporates ProtoCosmos Theory, HyperCosmos Theory, Limos Theory, Second Kind HyperCosmos Theory and HyperUnification Spaces. He has introduced PseudoFermion wave functions and theory, He has related Cosmos Theory to Regge trajectories of spaces, parton theory, Veneziano amplitudes, Fibonacci numbers and Ramsey numbers. He has calculated an approximation to the difficult R(n,n) Ramsey numbers.

He has developed a Gambol Model that successfully accounts for e-p deep inelastic scattering, fundamental particle resonances, hadron scattering, and the inner structure of particles based on confinement through Casimir forces of ideal gambol gases. The Gambol Planckian Distribution was derived.

He has applied the Gambol Model to particles, universes, and the Cosmos of universes. He showed that the Cosmos may have a distribution of 23 universes corresponding to various Cosmos spaces.

Recently he showed that Cosmos Theory follows from the number of independent asymmetric tensors in a dimension r. He also showed the close parallel between the form of $\gamma$-matrices and Cosmos Theory dimension arrays. The closeness suggested that dimension arrays have the same importance as $\gamma$-matrices for fermions.

He demonstrated that the pressure of fermions within a space of dimension r balances the Casimir vacuum energy force for 18 dimensions. He showed that $2e\pi = 17.02$ marks the critical point where pressure balances Casimir force, which implies r = 18 is the highest dimension Physical Cosmos space. The dimension $2e\pi$ appears to set the approximate dimension for Cosmos spaces with dimension array size $2^{r+4} \cong (17.02/8)^{r+4} . \cong (e\pi/4)^{r+4} . \cong 2.13^{r+4} .$

Now he has found the sequences of Coupling Constant values and fermion masses in the Standard Model and UST. This book unifies the Cosmos spaces spectrum, the coupling constant spectrum and the fermion mass sequences in a fundamental scaling differential equation.